Illuminated Revelations

THE NEAR DECLINE OF PHYSICS
DUE TO ITS UNDRESSED TERMS

The quarks, those constituents of the orgy
Playfully bound within the nucleons' chamber
Are named *up, down, strange, charm, bottom and top,*
The last two once being called *beauty,* and *truth*;

However, when just one of a type was contained
It became referred to, say, as a naked beauty,
And thus nude tops & bottoms their charms revealed—
To ever be in closeness binding, and bonding,

So, they even tried just *u, d, s, c, b,* and *t*
To prevent some ultimate collapse of physics,
But the truth of the flavors beneath the veils
Remained as the sheerest vision preferred.

So, we have these vibrant dancing ladies:
The naked heavyweight top, charming up,
And, down, the strange beauty of the raw truth,
With a bare bottom just around and behind.

They gyrate, spinning their charms, twirling,
In the universal dance of stunning motion,
The polarity sometimes reversed,
Whirling, their bottoms up and tops down.

And then there are Eden's many colors,
In this flower garden filled with flavors,
Such as red bottom beauties, blue tops,
And magenta undulations unstopped.

Illuminated Revelations

Gluons are the bees of the flower beds,
Carrying pollen back and forth to bond
The many relationships that make
This loved world go 'round as reality.

Eyed in views that probe the fundamental,
Quarks strangely swirl in and out of sight,
Pulsing, throbbing with elemental delight,
In and out—the love-made life of eternity.

These attractions in the altogether denuded
In the buff became the strong force, manifest,
That the mother-nature-naked terms exposed
To denote the stark beauty of truth uncovered.

"THREE QUARKS FOR MUSTER MARK"

Naked quarks would really love to go wild and dance,
But there's only a finite amount of energy and chance;
So, they would spiral out of control, having quite a blast!
Such, they have been confined within the proton—to last.

They're made bottoms-up;
Can we see them tops-down, a go-go?

No, for the quantum censor protects the charm show,
Their strange beauty and flavor bound up and down,
As the proton is much immune to disturbance around.

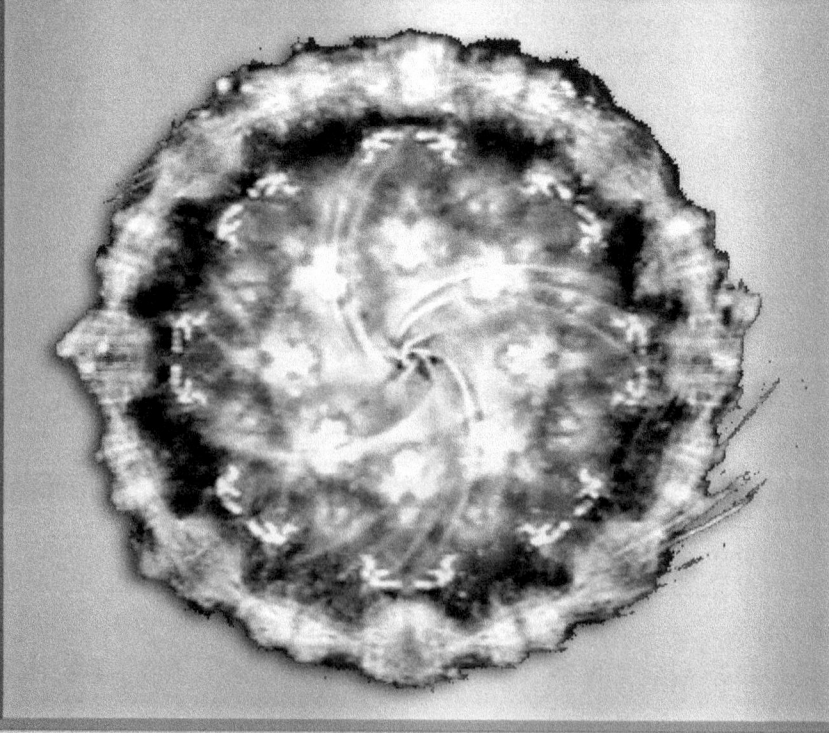

THE ENTRANCING DANCING

They were all dancing within love's treasure vault
Within the framework of a broadening thought,
The lights pulsing and the waves reverberating,
Where the good times had become everlasting.

Tribal primal field currents were raging
From speakers of the energy matrix pounding;
They whirled and twirled as loving gestalts
Of sentient consciousness knowing no halt.

There were rhythms of constant contraction
And expansions of bosom-energy projections
Converted to scalar waves of blinking attraction
As fission and fusion beckoned the connections...

...Ever forming in this Omni-sound emporium
Where tone waves vibrated in waves of creation.

ALCHEMY HAPPENS VIA RADIOACTIVITY
AND
HOW OLD CAN THE EARTH BE?

Through E=MCC we see that vast energy reserves
Are bound up in small amounts of matter, preserved.

Henri Becquerel carelessly left a packet of uranium salts
On a wrapped photographic plate in his drawer vault.

Some time later, he was surprised to discover that
The salts had burned a 'light' impression into it.
The salts were emitting rays of some sort, curiously,
So, he turned the matter over to Marie Curie, literally.

Madam Curie and her new husband Pierre, with glee,
Noted that the rocks poured out great amounts of energy,
But they never diminished in size or changed in any way.
They were converting mass into energy very efficiently.

They also found polonium and radium, and a Nobel prize,
Along with Becquerel, in 1903, Einstein yet on the rise.

Radioactive elements decayed into other elements,
Noted Ernest Rutherford and colleague Fredrick Soddy;
One day you had an atom of uranium that "bled",
And the very next day you had an atom of lead.

It always took the same amount of days.
For half of the sample to decay,
And so this steady reliable rate of decay
Could be used in kind of a clocking way.

Illuminated Revelations

Tick-tock, how old was it?

More than 700 millions years worth!
This age was way more
Than anyone had given the Earth.
(5 billion would be closer to the answer.)

He lectured one day,
Taking out a piece of radioactive pitchblende,
Showing it to aging Kelvin,
But Kelvin rejected it to the end.

Dimitri Mendedeyev rejected it too,
As with everything new,
Ever storming out of labs
And lecture halls all over, too;
However, the 101st element
Was called mendelevium,
In his name meant,
And quite appropriately,
For it was a very unstable element.

Pierre Curie began to experience
Radiation sickness, getting weak,
But in 1906 he was fatally run over
By a carriage on a Paris street.

Marie worked on with much distinction,
But had an affair so indiscreet
That even the French were scandalized there,
And so she was never elected
To the Academy of Sciences,
Despite not just one,
But two Nobel prizes
(Physics, Chemistry).

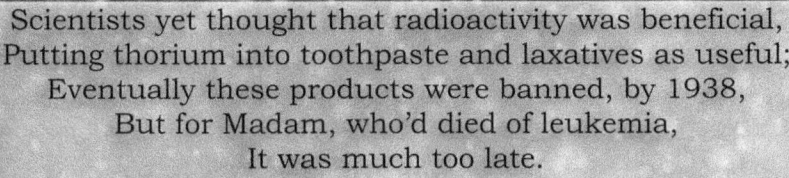

Scientists yet thought that radioactivity was beneficial,
Putting thorium into toothpaste and laxatives as useful;
Eventually these products were banned, by 1938,
But for Madam, who'd died of leukemia,
It was much too late.

The radiation is so pernicious and long lasting
That even now her papers from the 1890's,
And even her cookbooks, are dangerous and toxic,
So, all her lab books must be kept in lead lined boxes.
(One must wear protective clothing to look at them.)

Marie Curie was a very attractive lady, very much aglow,
For my great ancestor in his old writings such told me so.
She radiated warmth unto him as a rainbow of sparks—
"Great balls of fire!" he remarked,
"They now glow in the dark!"

Play all night long
with these

glow-in-the-dark
radioactive balls.

MOONLIGHT SONATA

The music of the spring was in the breeze,
A prelude borne by airy musicians
Of the trees—the mating calls of the birds,
That opened for the cosmic symphony.

The Music of the Spheres played in the park
At night—flung down by our Father, the Sky,
Through the soft night to our Mother, the Earth,
Then to us, their audience and progeny.

The planets joined in a concert to the
Merrie Monthe of Maie, arrayed as follows:
There was Venusia, the Bringer of Peace,
Singing side by side with warring Marsius.

Flitting about was the wingéd Mercuria,
The speedy messenger who conducted
The orchestra, melting all of us who
Were touched by her wand of burning desire.

And mighty Zeus, was there, full to the brim
With the jollity of the fat man's belly.
By Jove, came Saturnus, so very gray
With age—lumbering into the party.

Thence sat Urania, the magician, and
The old sea captain, King Nep, the mystic,
But not Pluto; he was downsized, no more
One of the harmonics—an underworld!

Illuminated Revelations

Jupiter's music was round and robust,
While Saturn's boomed with sounds of grandeur
And the old venerable melodies;
But, Mercury soon picked up the pace.

Next flowed the serene love songs of Venus,
Followed inexorably by Martial marches.
Now was the time for Urania's magic—
She played musical jokes and surprises.

At last, their music came to mesh as one,
And our wanderers of the night floated
Away on the haunting mystical strains
Of King Nep's tune, into the May Flower moon.

Now we're touched, so touched by the starlight,
Afraid that we'll ne'er be the same again.
Can you sense the euphony of the spheres?
Can you fathom the theory of everything?

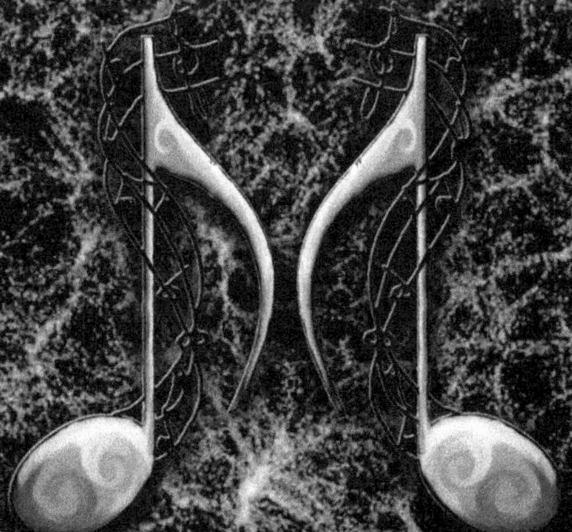

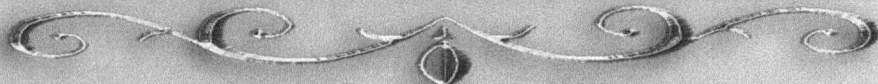

VACATION PLANETS

Uranus is quite pleasant compared to Pluto.

If you've ever had a dog, you know what I mean;
However, the under-worlded canine has been
Banished from the house of Astro—
To reign as the under-world in the Underworld,
For it's much better to reign in Hell
Than to be an unwelcome guest in the heavens.

Once, I was down on Venus,
And the sulfurous emanations
Were so repulsive that any gases from Uranus
Would have been to me as a breath of fresh air.

The gas giant planets' breadth and width is staggering,
And their mooning around is getting out of hand.

That leaves Mars as the only other good place—
Since Klingons have now appeared
On the rings around Uranus.

Illuminated Revelations

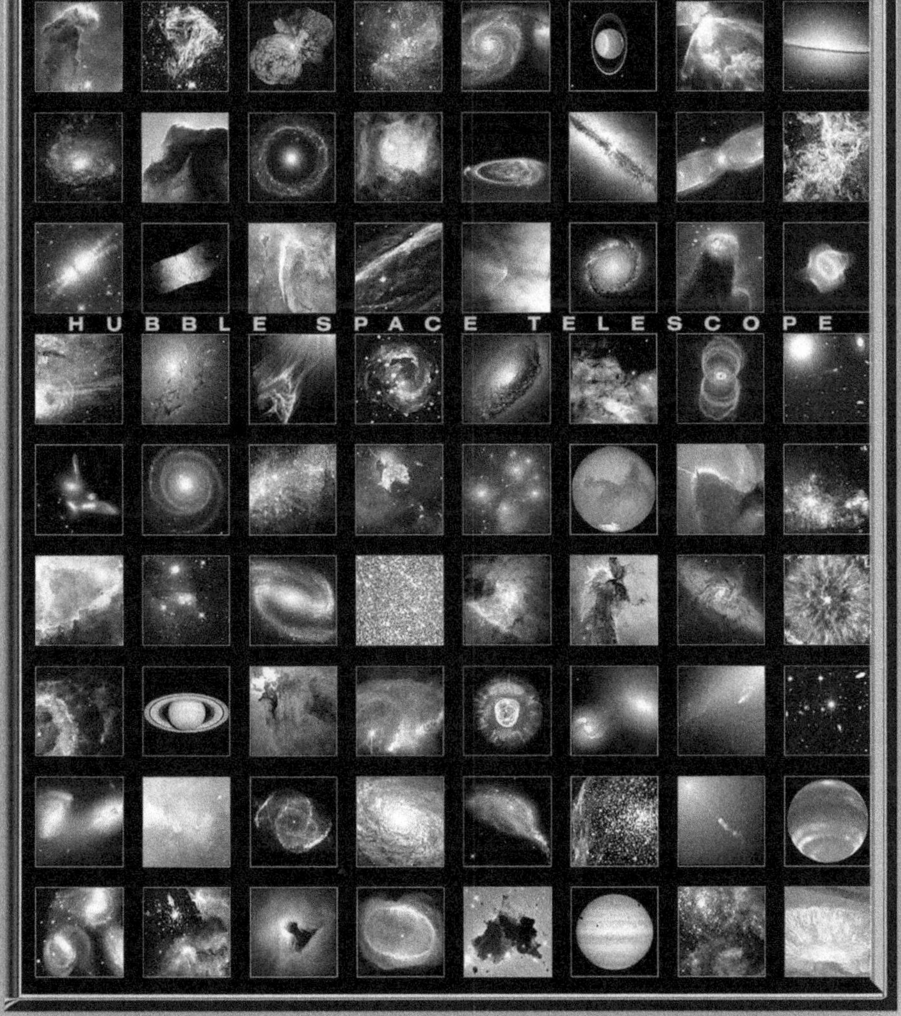

HUBBLE SPACE TELESCOPE

THE TIME CAPSULE

Since one million years had just passed by,
They, of the future, prepared to open, nigh,
The absolutely sealed container's prize,
Of a capsule made so carefully that it did survive
Without damage, being totally impregnable
To any outside influence imaginable.

They expected to see, perhaps, some old relic,
But certainly nothing alive that could tell of it,
For it would be hard to imagine, even then,
That some organism could keep on going its ken
Over its course of a million years later,
Sealed inside this tight container,
Unable even to exchange energy's spark,
This metabolism being the hallmark
Of life and all that quacked or quarked...

And, so, they did not at all expect something
In there that would be flapping its wings,
Gasping for air, or anything at all of life's song,
Wondering what had taken so long.

Well, they were right and they were wrong,
For in the time capsule that was planted so long,
Several things had with it come along...

One was a plaque, of numbers low and high,
And containing some primes and pi,
Another, some essays of the future—
Some, like Austin's, quite mature,

Along with maps and other items of the world
From those times when the oceans swirled;
But, the last, one perhaps not intended,
Was a microbe—an extremophile—
Laying there quite contented all the while!

Well, they soon laughed, loud and long,
For they were in between right and wrong
About what could survive from so long ago,
For, it was really walking mighty slow!

Amazingly, as with it the time capsule took,
The microbe walked right out of this book!

THE IRRESISTIBLE MEETS THE IMMOVABLE

She wanted to charge all those irresistible objects,
Even straight out of her husband's pockets;
But that grouchy and unlovable force
Said she could but window shop, of course.

So she bought stain-glass windows,
All very expensive ones, too,
For all the bedrooms and the baths,
A very large number, in her wrath.

She went to the hospital for the emergency
Of the credit card cut of plastic surgery;
'Twas nothing they could do for the charge,
Unable even to puzzle it together at large.

She returned home very much negative,
Finding him accounting all his positives.

"Stick 'em up," said she, assertively,
A gun pointed at his fatherlies;
"Give me your money or your life
If you want me as your loving wife!"

Said he, "Do I need all this rife and strife?
Go take up music and blow a fife!"

"When I was making the lion's fare,
All of that with you I did share.
I need to glamorously amorous
To keep our relationship harmonious."

He opened a drawer in his brain, that Mister,
That one of his mental cash register;
He then closed this manly box, in fact,
And into his brain he put it back.

So, on and on this all ever went,
But it was still that nothing could be spent.
She robbed his wallet while he slept
And left all the house very unkept.

Paternity and Maternity soon had it out
This battle an evenly matched bout;

The irresistible force met the immovable object,
Of which result no one could ever suspect.

What would happen, nobody knew;
And then everything really blew;

A total annihilation had occurred,
Nothing left at all but for a blur
A noisy bang, and a flash of light,
Such as with all creation's might.

All had been shredded and torn,
As when the universe was born.

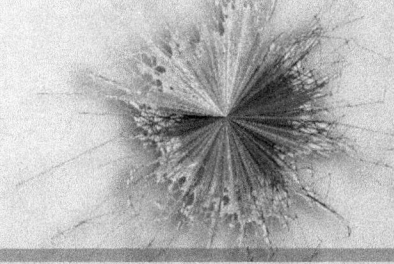

NOW AND ZEN

Everything that is part of us—
Our cells, tissues, organs and organ systems—
Has come about over billions of years
Because it proved successful
In the great survival stakes
During our perilous evolutionary
Descent (ascent) with modification.

The brain, being no exception,
Evolved, in part,
To allow a creature to learn
From what happens in its life,
To retain key elements that
Could influence future actions.

We are geared for self-preservation.
We will do anything to avoid facing the possibility
That who we are now cannot continue.

We ourselves are mainly the cause
That we are interested in.
The self is preoccupied with staying alive,
Which is why our species is still around today.

It is a prime biological function to be afraid of death,
And, so, the self, as thus contrived,
Is able to fully play its crucial survival role.

We want to equip our brain with a soul
That offers us an escape when the brain dies
Since the self cannot come to terms
With its own extinction.

Illuminated Revelations

Purest moonlight fell into the wrong hand,
As Evil swirled 'round—a drifting black sand

That drank the silvery beam from the cup—
Till the moon shone no more across the land.

Illuminated Revelations

From a subjective standpoint,
We are all born equal and undifferentiated
(Before that, 'we' were dead),
But, as mature selves we make a distinction
Between the individual and the surroundings.

Still, the brain keeps changing throughout life,
In a pattern of the shifting flux of its neurons;
We gain and lose memories and feelings,
Essentially creating a new person over and over again.

The self is thus not so rock solid as it seems.
These moment-to-moment changes differ from death
Only in degree. In essence, they are identical,
Although at the opposite ends of the spectrum.

So, we are not static things.
Other neural networks will come to be in other,
Future people, albeit with an "amnesia"
Of what went on before in
The brains of the previous others.

Why should we be happy about this?
We never can be, because the 'I' cannot operate
Outside of its own boundaries.
The only viable alternative is to think of a way
In which it is possible to ever continue on.

What will it be like to be a part
Of someone else after we die,
With our own particular
Narrative of life cast aside?

This is the 'zen' of now and then and when.

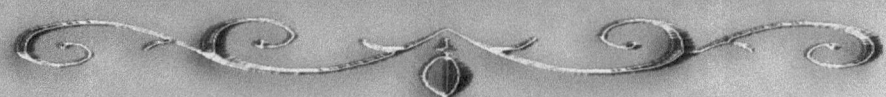

Illuminated Revelations

(When everything expires of that fate,
I'll still try to drink after the expiration date.)

THE FISH WHO ALMOST EVOLVED MORE

On the road to Kingston, NY, the other day
I was happy to see that the rains had returned
And that the drought was ending.

I stopped to do a little fishing along the way,
And it reminded me of a fish
I'd caught during the dry season.

I'll tell you about it now.
It took me awhile to get over it.

It was so dry that I could walk
Across the reservoir and the creeks.
The water was shallow due to the drought,
And most of the fish were swimming sideways
So they could stay under water.

Then I saw a rather amazing sight:
One fish was leaping from puddle to puddle,
Sometimes crawling across the dry land in between.

I threw away my fishing pole
And caught this fish with my bare hands,
Thinking of the delicious fish fry dinner
that I would have that night.

I put the fish in a bucket of water
In the front seat of my car
To keep it fresh during
The long drive home.

Illuminated Revelations

We dug the worms at night, keeping them moist,
And got up with the sun to fish, and then
Skinned them, and cooked them for lunch or dinner—
This to me is America Remembered.

Illuminated Revelations

Every so often that fish would poke its head
Out of the water bucket and look at me,
Sometimes even trying to jump out.

Finally, it did get out of the bucket
And sat on the seat next to me.
It was then that I realized that
I could never eat this fish.

About the same time, a brilliant idea stuck me:
I would train this fish to live out of water,
And make a pet out of it,
As the fish seemed to already
Have inclinations in that direction.

At home I put the fish in a barrel of water,
And sure enough, it tried to jump out.
So, each morning I would take it out
And put it on the grass,
Which was still wet from the dew.

Then, when I could see that it had had enough,
I would put it back in the barrel.
Each day the fish seemed to last longer and longer
Outside the water barrel before getting listless.

After a few months of this,
The fish didn't need much water at all;
As I walked along the road in the morning,
It would wriggle along beside me
In the wet grass in the shade.

Later, when the day became really hot,
I would give the fish a drink from my water jug.

Illuminated Revelations

After a few more months of training,
The fish was able to flop and sort of 'swim' along
Down the middle of dusty roads.
And when I offered it a drink, it refused!

We even went to the beach together;
Of course; only I went swimming—
The fish just laid on the sand, getting a tan
And enjoying the breeze.

One day it was over 120 degrees
And the fish just had to have a drink,
So I gave it a dry beer.

Other than that,
The fish never touched water anymore,
Having become a land animal.

What a lovable pet!
It slept with me, saw movies with me,
Went out to parties with me,
Chased down tennis balls
And brought them back to me,
Rode on the back of my bike, etc.,
We were inseparable!

But then a tragedy happened:
We were walking down
The road together one day,
And passing over an old bridge;
Suddenly my fish fell
Between some loose boards
And on down into
The creek below and drowned.

Everything...

...

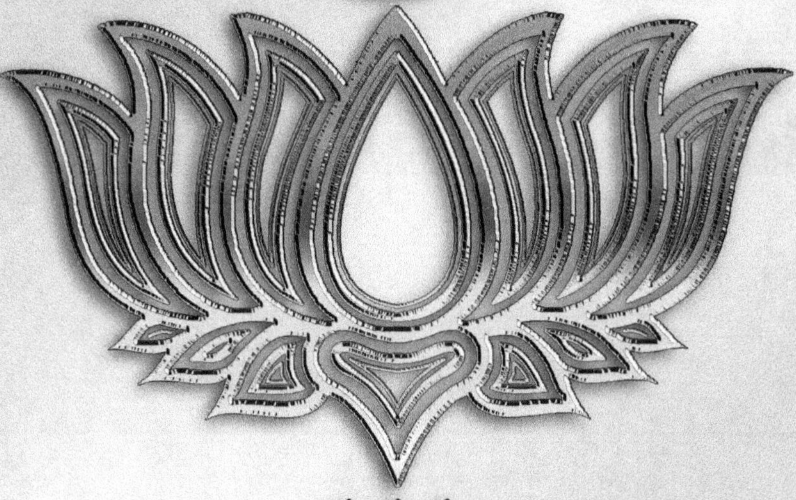

Watering Itself

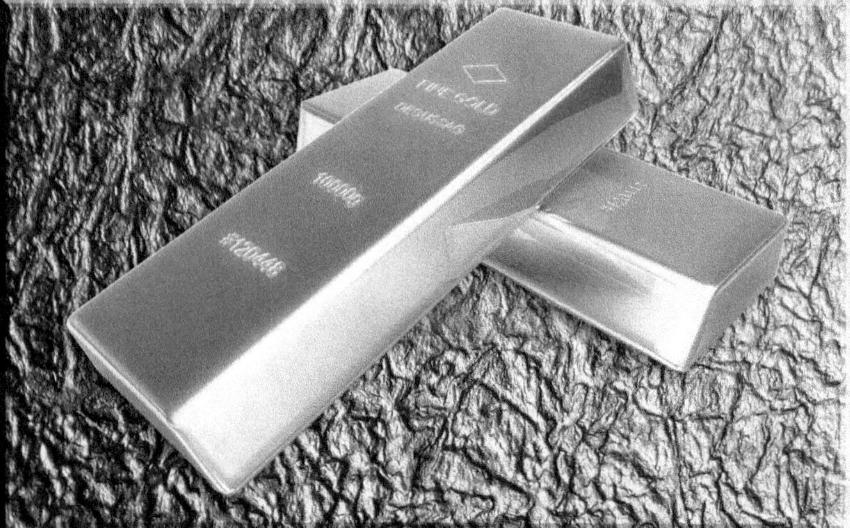

THE GOLDEN STREAM

In 1865, Hennig Brand thought that gold
Could be distilled from human urine, old,
Perhaps noting a similarity in color,
So, he kept fifty buckets in his cellar.

By some method, he converted urine
Into a noxious paste of some kind,
Then into some translucent waxy substance,
But so far there was no gold, and none hence.

However, after a time the substance began to glow,
And when exposed to the air burst as an inferno.
The substance soon became known as phosphorus,
But was too costly to make its business prosperous...

For an ounce of the flaming stuff sold
For way more than the price of gold!

Illuminated Revelations

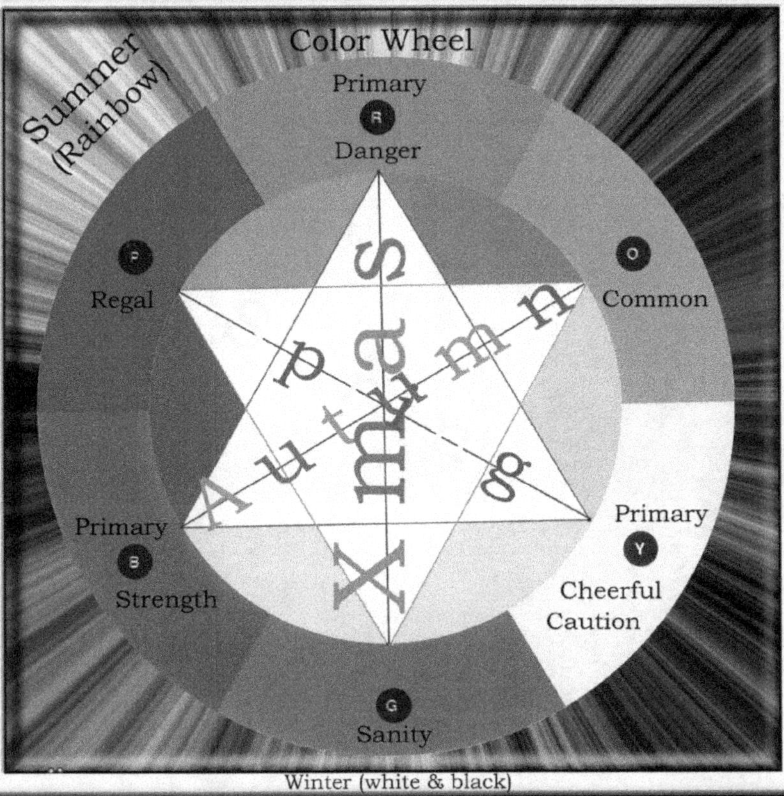

Color Wheel

Summer (Rainbow)

Primary
R
Danger

Regal
P

Common
O

Winter (white & black)

Primary
B
Strength

Sanity
G

Primary
Y
Cheerful Caution

Autumn

Strangely enough, all the color-pairs
That symbolize seasons and festive fairs
As they're found naturally in nature's ways,
Do contrast on the color wheel, crossways:

Illuminated Revelations

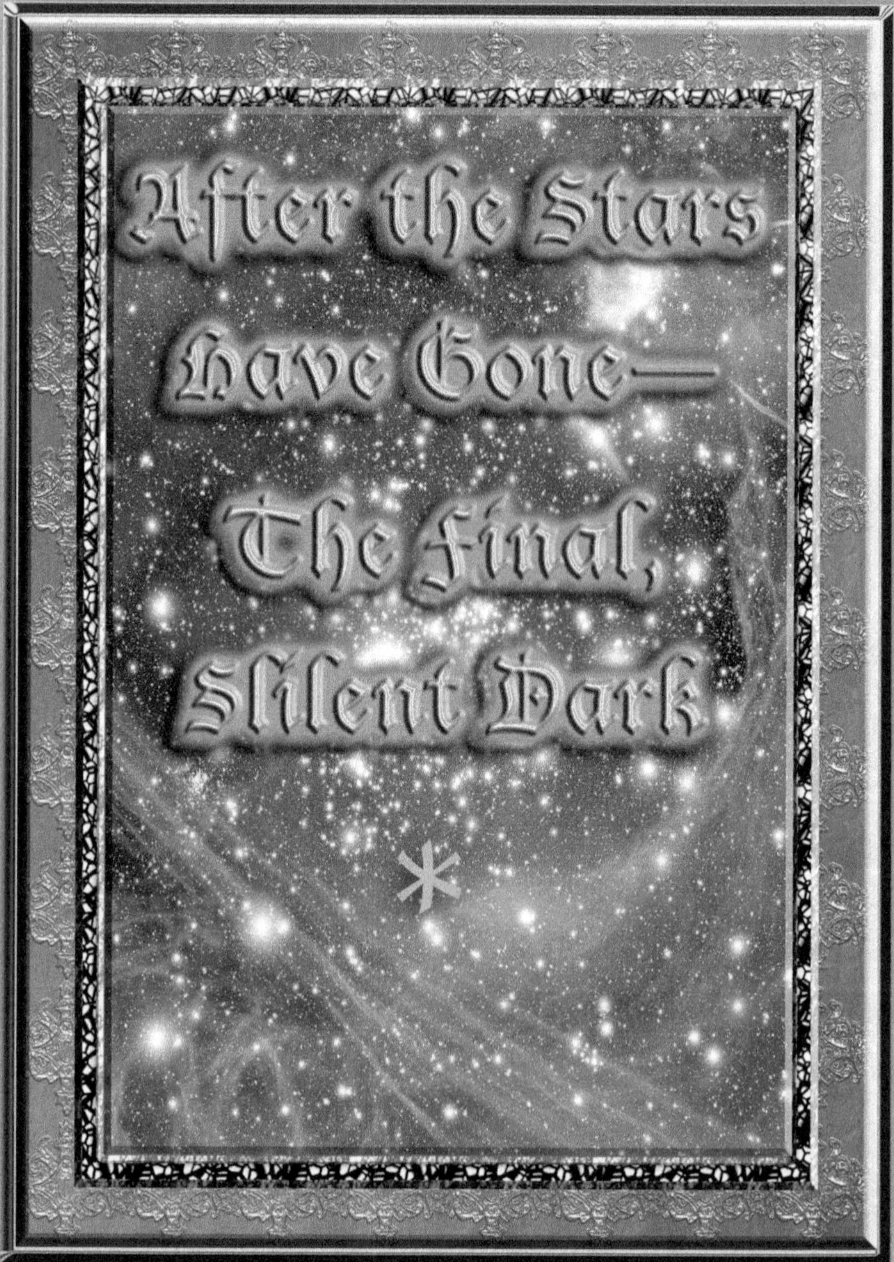

After the Stars Have Gone— The Final, Silent Dark

*

Illuminated Revelations

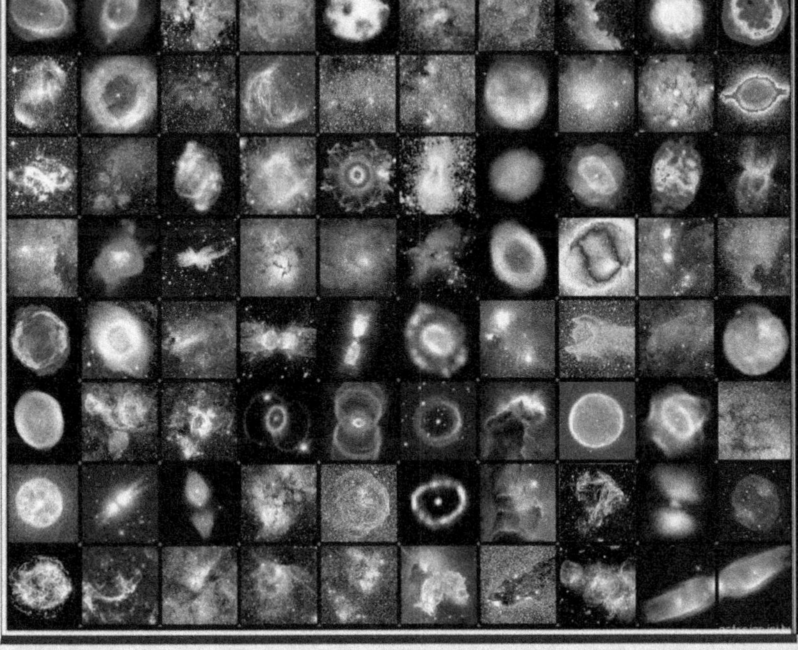

Nebulas

AFTER THE STARS HAVE GONE— THE FINAL, SILENT DARK

The Last Chance Saloon (Casino)

Entropy is always the winner in the end,
When there's no more money left to lend;
Meanwhile we stabilize, in nature's way,
Rearranging resources temporarily.

Prelude

Going beyond our very old obsession, so vast,
Of how it all began, back in the distant past,
But, retaining our search for meaning, from that,
We now turn to how will it all end, this and that,
Whether becoming collapsed, expended, or flat.

Is there is some deep meaning in all that?
Yes, for it is there in that future distance,
We'll find, or not, the end of our persistence—
Whether or not we are at all forever resistant;

Whether all that was, and what was did and done
Will be of any long-lasting benefit to anyone,
Of what destiny awaits, if there ever was one.

Endings are important to us, for what we're about,
Because we believe that how things turn out
Implies what the beginnings ultimately meant,
Of what, or not, is our place in the firmament.

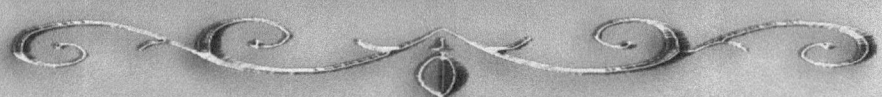

Illuminated Revelations

The stars are
Not just white,
They scintillate:

Sirius is blue,
Its companion green;

Betelgeuse, red;
Many, like Sol, yellow;

Arcturus, orange—
All jewels constellate.

Illuminated Revelations

As an ambitious species of nurture and nature,
We are now and always pointed toward the future,
For, of the three forms of the chimpanzee:
The common chimp, the bonobo, and us, we
Are the only chimp who went beyond the trees...

And, more importantly, even out of Africa, freed,
By that exodus, which laid down, indeed,
From that experience, the urge and the need
To move on, exploring, ever planting another seed.

The horizons on Earth sufficed us, as in "time",
For many millennia, but now the horizons' climes
Are broadened, through cosmology and physics,
And so they can well inform us of our prospects.

The future matters to us, for very basic reasons:
We wish to offset our mortality, our pleasin's,
To know if humanity's works, for every season,
Will be remembered, or lost; for nothing, even.

The Final, Silent Dark Marches On...

Time hurls a million waves of is displacement
At us, yet we are still there—our replacements:

Time, ever gray with age, hurls its changes, then,
'Gainst existence's rock, time and time again,
The entropic seas denuding the sands,
Yet, energy is preserved via science's wands.

Reminiscence weathered, but could ne'er wither,
For, in those mists of time; yesteryear yet appeared.

Illuminated Revelations

Illuminated Revelations

Would the prospect of a "Big Crunch" bring on phobia,
Such as an ever more confining claustrophobia?

Seems a better thought, somehow, though no picnic,
But more pleasing, if the universe(s) were also cyclic,
Although then all would still be really crushed
And forever lost, gone headlong into the rush.

We expect cycles, for all the days and seasons
Embedded this in our ancestors, into our reasons,
Since, at least, the periodic supplies some rhythm,
A pattern—the rolling hills of lives onward driven.

As for the cyclic, endless repetitions, they, too,
Would seem to revolt more of us than just a few;
As, too, perhaps, would some infinite abyss of time,
Which, too, grants us neither reason nor rhyme.

Does the drama go on forever, or does it end?
What do the visions of the future portend?
Doesn't it all have some purpose meant—
A goodly end of all of it to us might it present?
Is our higher mammal time, certainly,
But of such short parentheses within eternity?

It's only a finite time, then, which, too, tends
To horrify many, and more, as the universe ends,
Such as told by Robert Frost, a name of chill:
In heat or in cold, known as fire or ice, still.

Should we not believe in God since nothing lasts?
Well, if nothing lasts, then of what our purpose past?

Illuminated Revelations

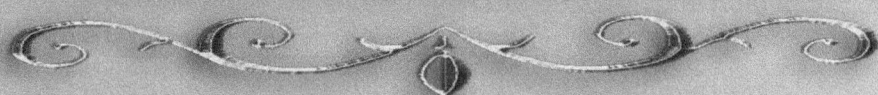

Illuminated Revelations

Is a purpose really required, so constructive,
Or would that be really quite restrictive?

No realm could really be special or sent,
Its becoming being of some specific intent,
For, all arrived here of causeless accident.

Is there anything wrong with the freedom to be,
Anywhere, any how, or any time during eternity?
No.

Should we rail against the law of entropy,
The "heat death" of thermodynamic energy,
The second of its final laws, we see,
Because it would destroy all of history?

Well, there are so many ways for disorder to be
Than any one ordered state specifically.

Would even a heaven on earth become a misery
If it, as it might, contain no more novelty?

Must there be an end to our revelry?
Can't we, at least, hibernate eternally?
Won't all matter, too, last eternally?

Will Shakespeare's works live on, paternally?
Is this not a Wagnerian struggle for eternity?

Science can settle whether a Last Day
Is ever going to come this way.

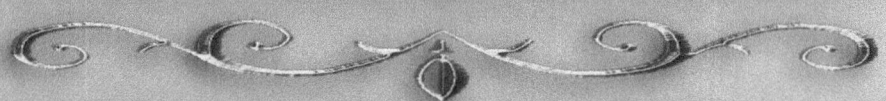

Illuminated Revelations

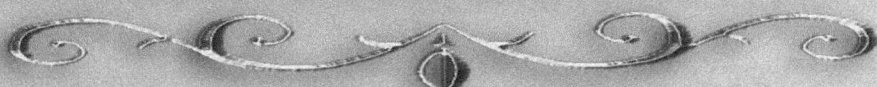

Illuminated Revelations

Only a decade or so ago, with some consternation,
We discovered the universe's large acceleration,
This expansion even increasing, onto some thin disaster,
The galaxies getting further away, ever and ever faster...
Then, one last snapshot taken, for all to remember.
The accelerating expansion of the universe's rafters
Means that the universe will cool even ever faster;
So, any conceivable forms of the future's life prolongers
Will have to keep themselves ever more cooler,
Think more slowly, and hibernate ever-longer.

One day the protons will fade away,
Leaving but dark matter, electrons, and positrons.
Everything was moving apart, cooling off,
The big slowdown not really so very far off;
Ultimately, even the black holes of late
And the lightless planets would dissipate.

The primordial soup, once so rich and hearty
Was now a thin gruel that couldn't serve the party.

One day, every particle would be moving away
From every other particle, so much out the way
That they won't even be able to see one another;
Thus, for all intents, motion will have ceased forever.

Our spurt of life, followed by an infinite stretch
Of dark equilibrium, was but the briefest sketch—
A warm and fuzzy stage, so interestingly active,
Whose time, relatively, was but infinitesimive.

Yet, we were there, in all our glory,
For whenever else could we be?

Illuminated Revelations

Illuminated Revelations

The Waves of the Ancient Swells
Of Time's Forgetting Tides
Swept Ever On...

As Time, now hoary with age,
Hurled forth its ashen change,
The charge ever san, pale and colorless,
That force born to summon decay, so endless,
'Gainst Nature's Universe each and every day;

Time and time again, Time fed all upon,
In its bloodless, white and waxen way;

But, this everlasting rose would not fade,
Its luster even brightening by the day,
Ever unsuccumbing to the sickly, peakèd
State draining drawn the life away;

Entropic seas yet denude the mountains,
Yet, this enduring flower, never-endingly
Has cast Deathly Time aside, for now,
Ceaselessly somehow thriving on,
To that which was the near imperishable,
The flame of beauty still inextinguishable,
Forever celebrated as immutable,
Gaining its seemingly perpetual permanence
From the undying love of the glorious truth.

Illuminated Revelations

Illuminated Revelations

In the future, uncounted societies of
Overlapping minds accumulate, with love,

In island redoubts, their preserved data burning
With a vital remembrance, in which, returning,
Past is the present and future, they all reliving
The data, even animating it and ever altering.

Without any new enrichments, the present and future
Reprise the past, in this retreat from external nature.

Their candles would have been nearly invisible to us,
They enduring, by diminishing, so as not to exhaust.

They made few new memories, a kind of blind sight,
For whatever realities had ever existed out of sight
Of their own mental structures were now fractured,
And thus not much different from those manufactured.

We sent message of early warnings to some,
In those castles of illusion, yes, many a one,
That they would face the decay, not so far away,
Of the heavy particles, the "proton pause", one day.

No self-assembled granularity can endure
Forever, but must return to the substructure,
And, so, the lives must all transition, it seems,
From heavier to much lighter regimes...

Although this, too, would not be permanent,
All destined to be swallowed by the firmament.

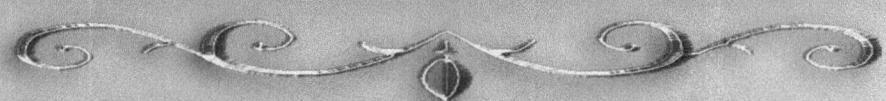

Illuminated Revelations

We sent message of early warnings to some,
In those castles of illusion, yes, many a one,
That they would face the decay, not so far away,
Of the heavy particles, the "proton pause", one day.

Illuminated Revelations

Every thousand years the Bird of Time
Flies over the mountain. A feather falls.
When the mountain has worn itself away,
The end of forever has thus arrived, that day.

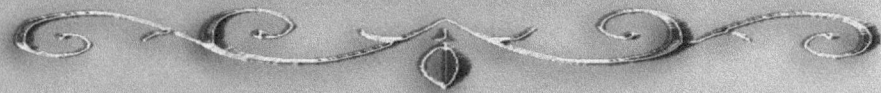

Illuminated Revelations

Illuminated Revelations

The Penultimate Part of the Final Dark

An Escalating One Way Trip
from a Fluke To Oblivion

The majority of the energy
Of the universe is dark today,
Although everything else passes
Through it in every way.

It's everywhere,
Having a component
That repels its own state,
Which cause the expansion of
The universe to much accelerate.

Dark Energy Matters:
The Escalation

We're on a one way trip from the quantum fluke,
That maximal energy within old Planck's nook—
Heading toward the oblivion of sparse expansion,
All that we ever loved and knew going to extinction.

Illuminated Revelations

Illuminated Revelations

We have often asked why some space exists,
Why it permits the countless to briefly persist
On Mother Earth nourished under Father Sky—
All of those finite sparks that light and die.

There were those who endlessly debated,
Whether to live in their virtuals unabated,
Or press forwards and outwards, of delirium,
To seek new localities in the mysterium.

It was much simpler once, in those days of old,
When we thought that universes didn't go cold,
But that they expanded and collapsed,
Still destroying all, yet ever giving more to last.
And, well before that, once upon a storied time,
We simply made it all up, with tales and rhyme.

The past was now a reef of dead accumulations,
A graveyard of various useless informations,
Which, despite their splendorous beauty,
Could not provide a novel futurity.

The last one of us, born of the sparkness,
Kept a window to the outer darkness…
S/he looked out, from a once brightly
Colored and sparkling inner reality,
Into the dark abyss…

There was nothing out there,
All being so lonely and bare—
No more singing of life's song;
For now everything was gone.

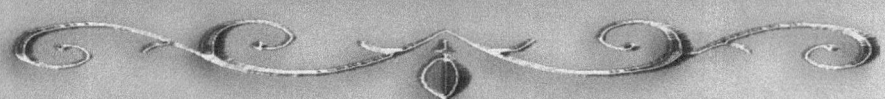

Illuminated Revelations

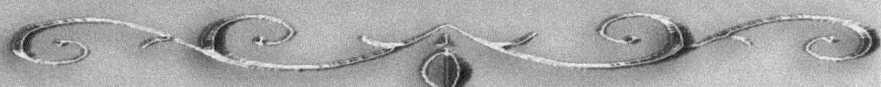

Illuminated Revelations

The Final Epilog

Multiple Verses

Our fruits are of a universal seed,
Are yet another yield of All possibility treed,
For siblings elsewhere in the entropic sea
Are also born of such probability.

There could not have been any special time,
One that was privileged over any other chime,
Nor any special place, nor any specific form
Arising out of the necessarily causeless realm.

Even those locally specific dates and places past
Of the events' novel memoirs could not ever last,
They being writ on water, with no meaning vast,
Disappearing in significance so very fast,
Since it's only the universals that last.

...

THE UNIVERSE

Illuminated Revelations

Illuminated Revelations

The protons were all gone from the show,
Having decayed so very long ago,
Into positrons—ever canceling the electrons,
But emitting the fleeing light of photons;
There being, of course, an equal amount
Of protons and electrons in the count;
And, of course, along with all the protons,
Went all of the atomic elements, the end,
All of their forms becoming myth and legend—
As they were still dreamt in night dreams,
Those forms that we once had, so it seemed.

S/he, as many of a luckily adaptable kind,
Had long since lightened and lighted the mind
With the dwindling electrons, and precious photons—
That beginning light of ancient times, growing wan.
Ours had been the only line in the uni-verse,
One that had become sentient, with proto-man first,
The rest of the cosmos being but a colossal waste,
A foreboding, harsh, and very dangerous place.

S/he was now the only one left,
Having outlived all of the rest.
The universe was near crumbling away,
Having run out of space, time, and all its sway.
S/he was dispersing, melting, into the vacuum, lone,
But, s/he held on for another thousand years, alone;
And, then, s/he, too, was gone,
The last of the hominid's song.

Illuminated Revelations

Illuminated Revelations

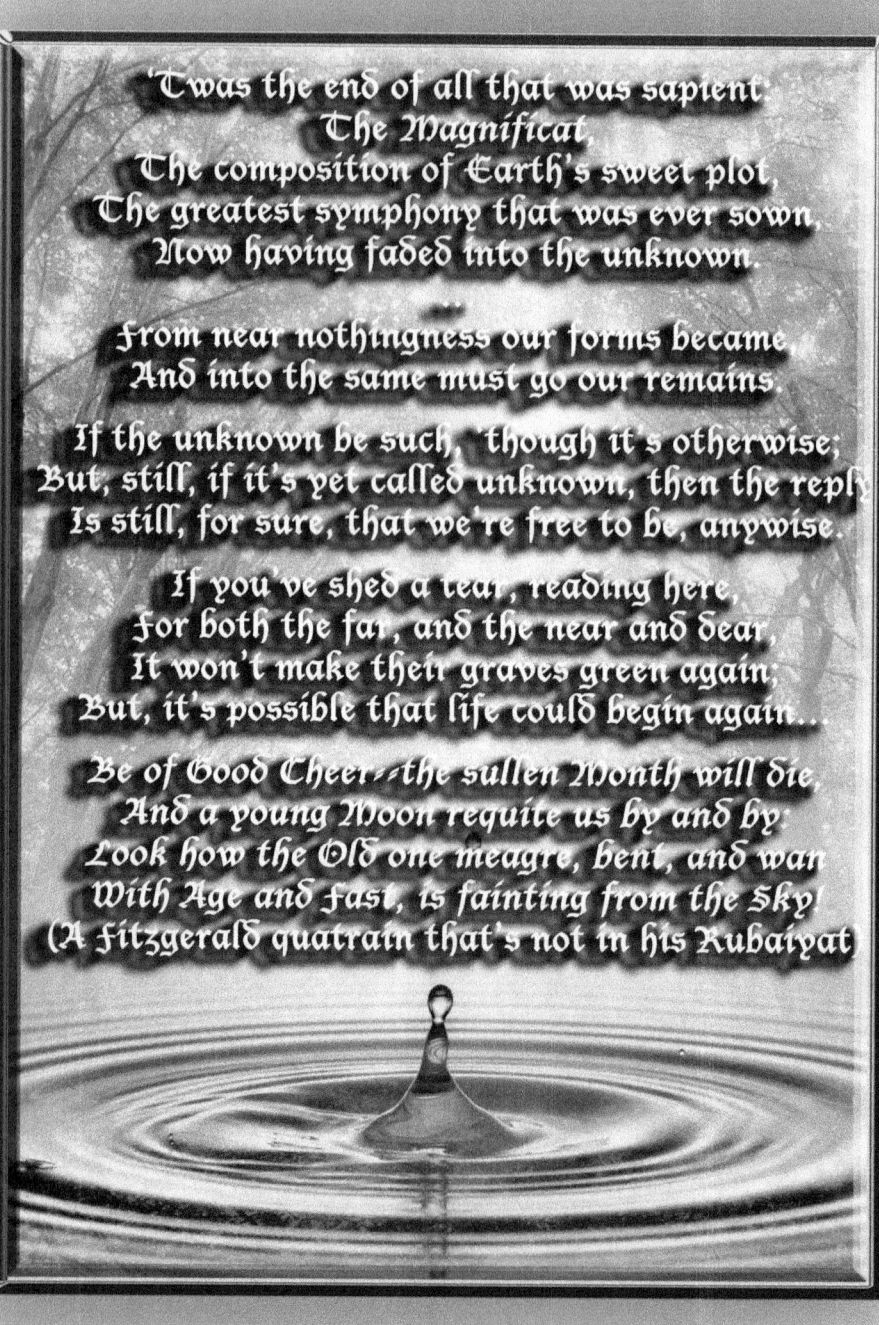

'Twas the end of all that was sapient;
The Magnificat,
The composition of Earth's sweet plot,
The greatest symphony that was ever sown,
Now having faded into the unknown.

From near nothingness our forms became,
And into the same must go our remains:

If the unknown be such, 'though it's otherwise;
But, still, if it's yet called unknown, then the reply
Is still, for sure, that we're free to be, anywise.

If you've shed a tear, reading here,
For both the far, and the near and dear,
It won't make their graves green again;
But, it's possible that life could begin again…

Be of Good Cheer—the sullen Month will die,
And a young Moon requite us by and by:
Look how the Old one meagre, bent, and wan
With Age and Fast, is fainting from the Sky!
(A Fitzgerald quatrain that's not in his Rubaiyat)

Illuminated Revelations

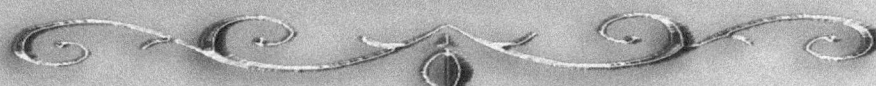

The Eternal Return

Behind the Veil, being that which ev'r thrives,
The Eternal Multi-Cycle has ever been alive.

Some time it needed to learn Everything for,
And now well knows how these bubbles to pour,
Of existence in some meant universe,
Those that wrote your poem and mine, every verse.

So, as thus, thou lives on yester's credit line,
In nowhere's midst—now in this life of thine,
As of its bowl our cup of brew was mixed
Into this state of being that's called "mine".

Yet worry you that this Cosmos is the last,
That the likes of us will become the past,
Space wondering whither whence we went
After the last of us her life has spent?

The Eternal Saki has thus formed
Trillions of baubles like ours, and will form,
Forevermore—the comings and passings
Of which it ever emits to immerse
In those universal bubbles blown and burst.

So, fear not that a debit close your
Account and mine, knowing the like no more;
The Eternal Cycle from its pot has pour'd
Zillions of bubbles like ours, and will pour.

When You and I behind the cloak are past,
But the long while the next universe shall last,
Which of one's approach and departure it grasps
As might the sea's self heed a pebble-cast.

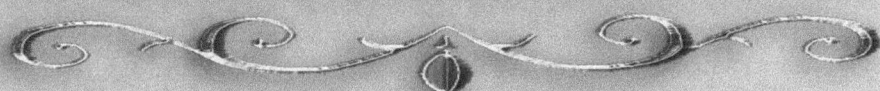

— Forever After —

When the sun burns out,
And, soon after,
When the Earth grows cold
From that disaster,
When galaxies die
And rotate no more—
Then what remains
Is our love, thereafter.

Illuminated Revelations

Illuminated Revelations

All That Lies Between

Illuminated Revelations

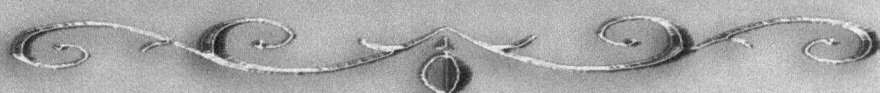

Illuminated Revelations

It is a beauty and a brilliance
flashing up in its destructance;
for, everything isn't here to stay its "best";
It's merely there to die in its sublimeness.

Like slow fires making their brands, it breeds;
Yet, ever consumes and moves on, as more it feeds,
Then spreads forth anew, this unpurposed dispersion,
An inexorable emergence with little reversion,
Ever becoming of its glorious excursions
Through the change that patient time restrains,
It feasting upon the glorious decayed remains,
In its progressive march through losses for gains.

Illuminated Revelations

Illuminated Revelations

We have oft described the causeless—
That which was always never the less,
As well as the beginnings of our quest,
And, too, have detailed, in the rarest of glimpses,
The slowing end of all of "forever's" chances;

So, then, we must now turn our attention keen
To all of the action that's here in-between—
All that's going on, and has gone before,
Out to the furthest reaches "ever-more",
For, everything that ever happens,
Including life and all our questions—
Meaning every single event ever gone on
Of both the animate and the non—
Is but from a single theme played upon.

This, then, is of the simplest analysis of all,
For it heeds mainly just one call—
That of the second law: dispersion,
The means for each and every occasion,
From the closest to the farthest range—
That which makes anything change.

These changes range from the simple,
Such as a bouncing ball resting still
Or ice melting that gives up its chill,
To the more complex, such as digestion,
Growth, death, and even reproduction.

Illuminated Revelations

Illuminated Revelations

There is excessively subtle change, as well,
Such as the formations of opinions tell
And the creation or rejections of the will..

And, yet, all these kinds of changes, of course,
Still become of one simple, common source,
Which is the underlying collapse into chaos—
The destiny of energy's unmotivated non purpose.

All that appears to us to be motive and purpose
Is in fact ultimately motiveless, without purpose;
Even aspirations and their achievement's ways
Have fed on, and come about through, the decay.

The deepest structure of change is but decay;
Although, it's not the quantity of energy's say
That causes decay, but the *quality*, for it strays.

Energy that is localized is potent to effect change,
And, in the course of causing change, it ranges,
Spreading, and becoming chaotically distributed,
Losing its *quality* but never of its quantity rid.

The key to all this, as we will see,
Is that it goes though stages wee,
And so it doesn't disperse all at once
As might one's paycheck before a month.

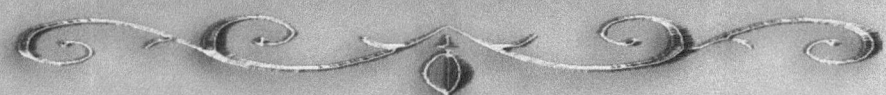

Illuminated Revelations

Illuminated Revelations

This harnessed decay results not only for
Civilizations, but for all the events going fore
In the world and the universe beyond,
It accounting for all discernible change,
Of all that ever gets so rearranged;
For, the quality of all this energy kinged
Declines, the universe unwinding, as a spring.

Chaos may temporarily recede,
Quality building up for a need,
As when cathedrals are built, or forms,
And when symphonies are performed;
But, these are but local deceits,
Born of our own conceits;
For, deeper in the world of kinds
The spring inescapably unwinds,
Driving its energy away—
As ALL is being driven by decay.

Illuminated Revelations

Illuminated Revelations

The quality of energy meant
Is of its dispersal's extent;
When it is totally precipitate,
It destroys; but when it's gait
Is geared through chains of events
It can produce civilization's tenants.

Ultimately, energy naturally
Spontaneously, and chaotically
Disperses, causing change, irreversibly;
Think of a crowd of atoms jostling
At first as a vigorous motion happening
In some corner of the atomic crowd;
They hand on their energy, loud
Inducing close neighbors to jostle, too
And soon the jostling disperses, too —
The irreversible change but the potion
Of the random, motiveless motions.

And such does hot metal cool, as atoms swirl
There being so many atoms in the world
Outside it than in the block metal itself;
Entropy's statisticals average themself.

Illuminated Revelations

Illuminated Revelations

The illusions of purpose lead us to think
That there are reasons, of some motive link,
Why one change occurs and not another;
And even that there are reasons that cover
Specific changes in locations of energy,
The energy choosing to go there, intentionally,
Such as a purpose for a change in structure,
This being as such as the opening of a flower;
Yet, this should not be confused with energy
Achieving to be there, in that specific bower,
Since, at root, of all the "power",
Even that of the root of the flower,
That there is, is the degradation by dispersal,
This being mostly non reversible, and universal.

The energy is always still spreading, thencely,
Even to some temporarily located density—
An illusion of specific change
In some region rearranged;

But, it's just lingering there, "discovering",
Until new opportunities arise for "exploring",
The consequences but of random opportunity,
Beneath which, purpose still vanishes entirely.

Illuminated Revelations

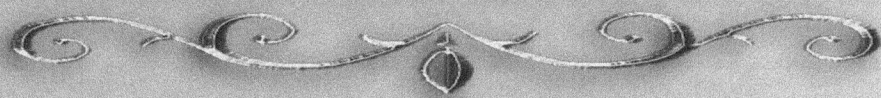

Illuminated Revelations

Events are the manifestations
Of overriding probability's instantiations—
Of all of the events of nature, of every sod,
From the bouncing ball to conceptions of gods,
Of even free will, evolution, and all ambition;
For, they're of our simple idea's elaborations;
Although, for the latter stated there
And such for that as warfare,
Their intrinsic simplicity
Is buried more deeply.

And yet, though sometimes concealed away,
The spring of all creation is just decay,
The consequence and "instruction"
Of the natural tendency to corruption.

Love or war become as factions
Through the agency of chemical reactions,
All actions being the chains of reactions,
Whether thinking, doing, or rapt in attention,
For all is of chemical reaction.

THE ONLY PURPOSE OF LIFE
IS TO BE—FINDING
YOUR OWN MEANING
THEREIN;

BUT,

SOME QUESTIONS
STILL REMAIN,

SUCH AS

"WHAT IS LIFE?"
(AND IT'S POINT).

TO FIND THE ANSWER,
ONE MUST LIVE IT FULLY!
(WITH GOODNESS)

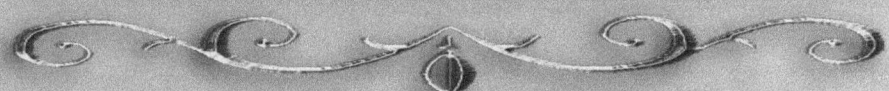

Illuminated Revelations

At its most rudimentary bottom,
Chemical reactions are rearrangement of atoms,
These being species of molecules,
That, with perhaps additions and deletions
Then go on to constitute another one, by fate,
Although, they sometimes only change shape,
But, too, can be consumed and torn apart,
Either as a whole or in part, so cruel,
A source of atoms for another molecule.
Molecules have neither motive nor purpose to act—
Neither an inclination to go on to react
Nor any urge to remain unreacted;
So, then, why do reactions occur, if unacted?

Molecules are but loosely structured
And so they can be easily ruptured,
For reactions may occur if the process energy norm
Is degraded into a more dispersed and chaotic form,
And, so, as they usually are always constantly subject
To the tendency to lose energy as the "abject"
Jostling carries it away to the surroundations,
Reactions being misadventure's transformations,
It then being that some transient arrangements
May suddenly be "frozen" into "permanences"
As the energy leaps away to other "experiences".

To future columns
We stretch our present row,
By a lifeline
Of tenuously spun vow.
Oh how soon the weighted web
Begins to fail—
The only real time
Under our feet is NOW.

Illuminated Revelations

So, molecules are a stage in which the play goes on—
But not so fast that the forms cannot seize upon;
But, really, why do molecules have such fragility,
For, if their atoms were as tightly bound as nuclei,
Then the universe would have died, being frozen,
Long before the awakening the forms "chosen",
Or, if molecules were as totally free to react
Every single time they touched a neighbor's pact
Then all events would have taken place so rapidly
And so very crazily and haphazardly
That the rich attributes of the world we know
Would not have had the time to grow

Ah, but is it all of the necessitated restraint
For it ever takes time the scene to paint
As such as in the unfolding of a leaf—
The endurations for any stepping feat,
As of the emergence of consciousness
And the paused ends of energy's restlessness:
Is of the controlled consequence of collapse
Rather than one that's wholly precipitous.

So, now all is known, of our here's and nows
Within this parentheses of the eternal bough,
As well as the why and how of it all has come,
And of our universe's end—but, that others become.

Illuminated Revelations

(The verse lines, being like molecules, warmed,
 Continually broke apart and reformed
About the rhymes which tried to be nonintrusions,
 Eventually all flexibly stabilizing to conclusion.)

Illuminated Revelations

The Symphony of Life

All that we know
Even the lovliest of the best,

Decomposes into the dust
Of earth compressed.

The songs once composed
Now lie in repose;

Of this dust the future
Arrranges and recomposes.

Out of energy's dispersions and decay
Comes growth and the emergence of complexity.

Illuminated Revelations

The End ...
and
The New Beginning

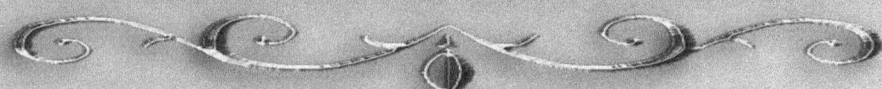

SIMPLICITY

Occam sharpened his razor,
To a one-dimensional line,
Then cut his beard into strings.

They sprung from the depths,
Vibrating the songs of reality,
For which all composites sprang.

OCCAM'S ALPHABET

—BCD—FGH—KLN—P—RST—VW—Y—

In the alphabet, Occam saw the unnecessary,
So he struck out 'j', 'q', 'x", and 'z',
Being rarities or duplicates,
And then even cut more,
Those being the vowels taking up space.

n th lphbt, ccm sw th nncssry,
S h strck t ", ", ", and ",
Bng rrts r dplcts,
nd thn vn ct mr,
Ths bng th vwls tkng p spc.

But then one could only understand him almost.
Bt thn n cld nly ndrstnd hm lmst.

CONCISE SIMPLICITY

Writers of few words,
Even the laws that writ reality,
Can often say more more with less.

<u>WHAT SENSE TO MAKE?</u>
<u>WHICH PATH TO TAKE?</u>

From what beastly heart
Springs our zest?
Of what searching eye
Became our sight?
What sound in the bushes
Let us hear?
What dark past haunts
But helps us be?

Across what ink black river
Did we have to swim?
To what ends at length
Did we search for food?
In what deep entangled forest
Were we bred?

And hitherto,
Of what stars did we shine of their stead?
And in what nursery were those infants fed

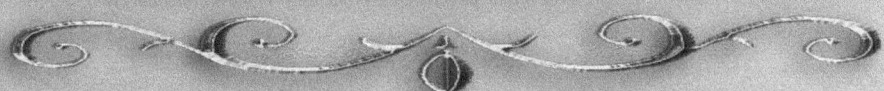

THE VAULT OF EVERYTHING

A spirit led us onward, who knows how,
Toward the Library of Babel,
Which contains all the possible books
That could ever be written,
Including, for example,
Better and worse Shakespeare plays,
Brand new plays,
Books with only one word
Of difference among them,
Everyone's life story
(Even the parts not lived yet),
The Secrets of the Universe,
The true Theory of Everything,
A lot of gibberish, and so on,
As we can't imagine.

[In fact, I found this story in there,
In a short story book of mine-to-be,
So I just copied it to here.
(Yes, it said that too.)]

A clear night sky of infinite possibility
Showered us with photons,
Lighting our way
To the fountain of all knowledge.

"True enlightenment awaits me there,"
I offered to the guiding spirit.

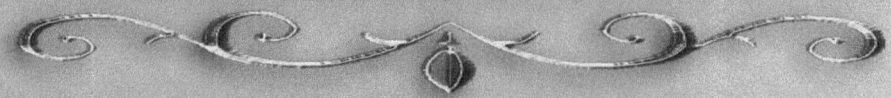

Illuminated Revelations

The Vault of Everything

"Don't be so sure,
Although you might chance upon it,
For the deep truths of enlightenment
Are as needles surrounded and consumed
By the near infinities of the stacks
Of deception and confusion,
For, remember,
EVERYTHING exists in this library."

"It must be a massive building," I remarked.

"Well, yes, but it's bigger on the inside
Than on the outside;
Otherwise, it would have been
Larger than the universe."

"Bigger on the inside? How?"

"Well, you'll see, but I'm not sure how—
Maybe through some dimensional extensions—
Or perhaps it's constructed digitally
And expands as you move about, somehow,
To conserve space;
But, even with compression,
It's still hundreds of miles wide
In every direction—on the inside."

"What is Everything, in principle?"

"Every arrangement possible,
Given whatever constraints there are, if any.
Of course, not all paths may be stable,
Sensible, or last very long."

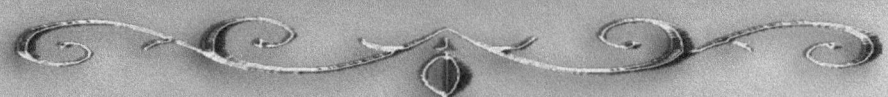

Illuminated Revelations

Earth could not answer, nor the Seas that mourn
In flowing Purple, of their Lord forlorn

Nor rolling Heaven, with all his Signs reveal'd
And hidden by the sleeve of Night and Morn.

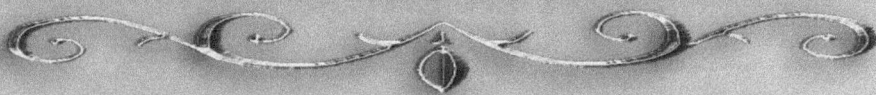

"That's a lot—
Why do we live on this particular path
That our Universe has taken?"

"Who the heck knows!"

"What about making the forms of
Substance(s) of a Universe?"

*"Well, in the case of the emission
Of the secondary substance(s), let's say,
It's every one of the 'alphabets'
That can be conceived by
The Timeless-Formless-Motionless,
Plus, all of its resultant
Workable combinations
And interactions of substance.*

*For this Babel library,
It is every possible arrangement
Of words in every language,
With punctuation, too, naturally."*

"Hey, here it is. I can't wait!"

Upon entering, they saw stacks of books
In every direction, even up and down,
Stretching toward infinity.

"Where's the card catalog?"

*"There can't be any,
For many titles and descriptions
Of similar books are too long to differentiate.*

Illuminated Revelations

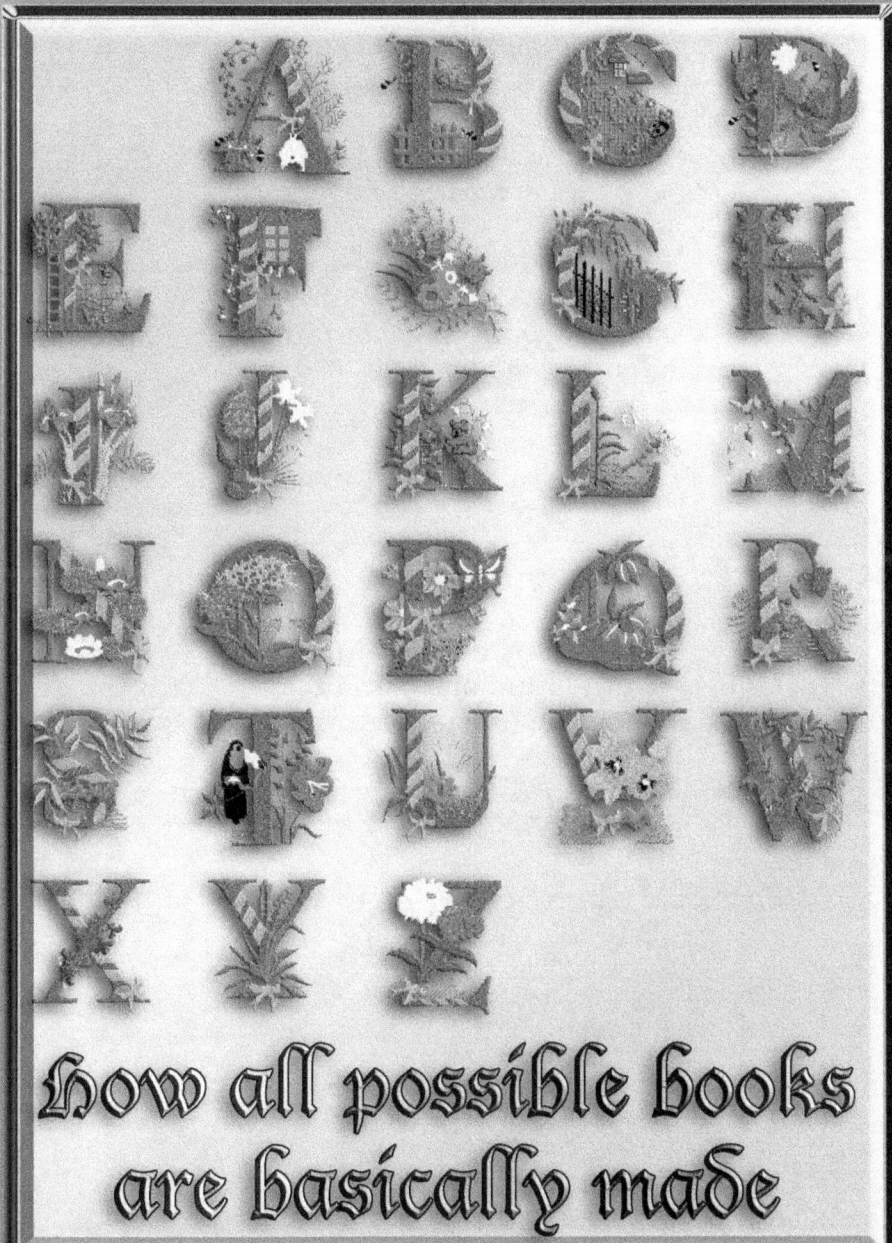

how all possible books are basically made

Illuminated Revelations

Think of the books themselves
As the card catalog."

"How's the library organized?"
"It can't be. It would take forever."

"Who runs it?"

"Borges is the lone librarian,
But he's somewhere in the back
And hasn't been seen for decades."

"OK, I'll pick some at random."
(Hours pass)

"Anything?"

"No, mostly mumbo-jumbo,
But I found one on a table
That someone must have treasured."

"Oh, yes, he spent his entire lifetime here.
It's Plato's 'Beyond Metaphysics'."

"Wow! That's been lost for thousands of years.
But is it the true version?"

"Who knows."

"This library contains
No information whatsoever!"

"True, but there's another library next door
That also claims to have Everything."

Illuminated Revelations

"You mean that little 'hut'
No, wait—I get it—
The library next door is empty."
"Yes, for the All sums to the None."

"Wait, I found two more good ones
In the stack right near the entrance…"

*"One is by you and one is by your friend, Rascal.
You put those there in the first stack
So someone would find them easily
And read them, even though they exist again
Somewhere else in the library."*

"Yes, and I'm even going to let them
Stick out a little on the shelf."
…
In another chilling Borges's story,
I read the actual book that he refers to,
The one whose infinite pages
Are ever-changing,
For that's how books appear to me
In my night dreams.

Sometimes there are
Even digits occurring
In the middle of words,
Plus, if I look away and then back,
Then the contents of the page have changed.

One time, when the page stabilized
To quite understandable words,
I realized I was reading
Something very profound.

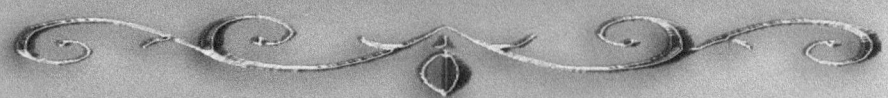

Illuminated Revelations

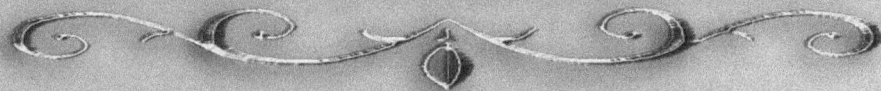

In fact, it was the Ultimate Answer.
I dared not look away
Nor try to copy it with a dream pencil,
But, instead, tore out the page
And crumbled it into my hand,
Then forced myself awake
(it was a lucid dream).

When I awoke,
I had the page in my hand,
And it said:

This page intentionally left blank,
Except for the above,
And the above, etc.

Illuminated Revelations

GRACIOUSLY WELCOMING
LADY LUCK BECOMING

He believed that luck would never fail—
So he ran like the wind through the jungle,
Surely knowing. He'd what he'd come for,
Now hopeful to find the help at the shore.

The relentless ones were not far behind,
That ill-fated menace of the bad kind.

Miss Fortune laughed, and said,
"No road could be too hard to tread
For we are fearless. To those, a boon—
For they ever seize the Opportune."

"I see you, Fairest Happening."

Just past a sharp turn, in the trees,
He suddenly dropped to his knees
And fired into his pursuers mean
As they came upon the scene,
Using all his ammo but for one round,
Then hurried on with nary a sound.

"I am wide aware," Miss Karma,
"Of this continuing Dharma—
That chance shines as my sun,
For, she, in turn, happens on everyone."

"Oh, say it is your lot, my friend and lover,"
She answered back, granting him cover.

Illuminated Revelations

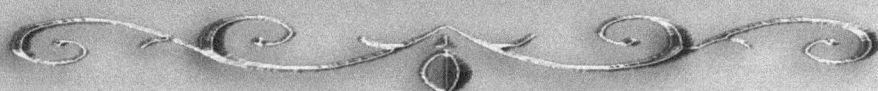

Listening, he could hear ever more troops
Rushing through the night in groups,
About a half-mile back around the loops.

"I gratefully welcome thee,
Miss Lady Luck of Dice,
Though I may pay a late fee
For my pick up so precise."

Ms. Destiny Serendipity smiled, saying,
"The game is on; we are playing.
Let joy and innocence prevail;
Believe that luck will never fail."

He moved on, ever faster, cheating death,
A third wind becoming of her vaporous breath,
It blowing this DIA operative onward
To the shore ever toward.

He could hear the whirling chopper,
But now receding was its Doppler,
He thus grieving
Of its leaving.

"Am I much too late—still too far?
Shall I curse you all, destined stars?"

"No," said lovely dear Twist of Fate,
For you have one bullet left for chance,
Not to use to sleep or dream perchance."

But the chopper was rising high,
Well into the star-crossed sky.

Illuminated Revelations

"Shall to self I take this bullet
Now that the bus has left?"

"Oh, no," Miss Lucky Break encouraged,
"Do not be at all discouraged,
For you know it shall not be so
And what with it you now must do."

"Yes, perhaps it shall be so in some plight
Coinciding in a most kempt and hapful night."

He smiled and then knelt to ground,
And sent his last bright tracer round
Just ahead of the copter now departing,
His minor wounds yet sorely smarting.

"I bless you with all my lucky charms,
My good and well-fated man of arms."

The door-gunner noted the red tracer
And whence it came of the river vapors.
"Captain, turn back and take a look;
He awaits a fortuitous accidental fluke."

"I am an uncursed, non-jinxed agent man.
Let my joyous innocence prevail again."

He jumped into the rescue's hovering haven,
Directing the door-gunner's firings, wavin'.

"Fare thee well, my nightly knight"
Dame Fortune wished upon his sight.
"You recognized me even in the dark."

Illuminated Revelations

Illuminated Revelations

"Oh, My Angel, Passiona, lovely lark,
I might have known it was you
That would ever see me through."

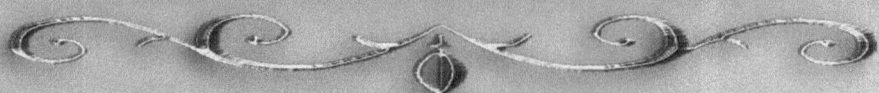

RAINBOW

Toward the end of a sunny day,
A storm came and washed away,
And the sunset clouds, being glad,
Held a party for the returning lad.

The sun then peeked, and soft shone
Into the mist of the departing squall,
Its light split into particolors lone,
Separating, each from the ALL—

A bouquet of colored rays
Swirled into sight,
And promised good weather
For the rest of the night.

The rainbow lit up the east,
As long we attended the feast
Of both the east and the west,
Till into darkness we descended blest.

THE END OF THE EARTH

The Asphodel sustains the Dis dwellers,
Where they rest beyond that fatal river—
There the wretched shades drink forgetfulness,
And to oblivion sink without distress.

Fireweed grows from Hell's sulfurous embers,
As does Purple Loosestrife—dead men's fingers;
But wildflower air revives the dead—and then
Those happy souls can thrive on Earth again.

Charon was withered, wan, and skeletal,
Although eternally grateful for his immortal life
And steady job of ferrying the dead across the river Styx
In their transition from life to death to forgetfulness.

As Earth was the only planet he'd come across
With such promising higher life forms,
Charon had grown rather fond of its inhabitants,
Even though he only saw but the worst of them;
But, even from this he could extrapolate
To the qualities of the best.

Charon did his job well, professionally,
Although it was ever so dreary
With the endless darkness of wasted lives
And the grim and gloomy skies all around,
For this land always had
That same gray and leaden feel.

He ferried on, though,
For his own life was precious to him.

The soon-to-be really really dead never said much,
For what was there to tell after an empty life
That had often turned to deep regret;

So, Charon did not prompt them for information,
For this was not the thing to do
At the time of their passing,
So he was always most
Courteous and kind to them,
Even to the most evil of the darkest,
Doing his task as well as he could.

It was not that Charon was afraid that
His undersized master of the underworld,
Pluto, might be watching,
But that he had the extreme clarity
To duly serve the task at hand,
A testament to his character.

Charon had been quite alarmed lately—
What with the numbers of the hellish-souls-to-be
Climbing into the millions in such a short time,
But, he had been through this kind of rush before,
With the doomed and damned of other planets
That had been consumed by their suns
Or had undergone other such catastrophes.

He just used larger boats
And patiently took his time,
For he had all of Eternity.

Of course,
Charon could and did feel deep sadness,
But he didn't show it outwardly,
Even when the numbers from Earth
Increased a thousand-fold again.

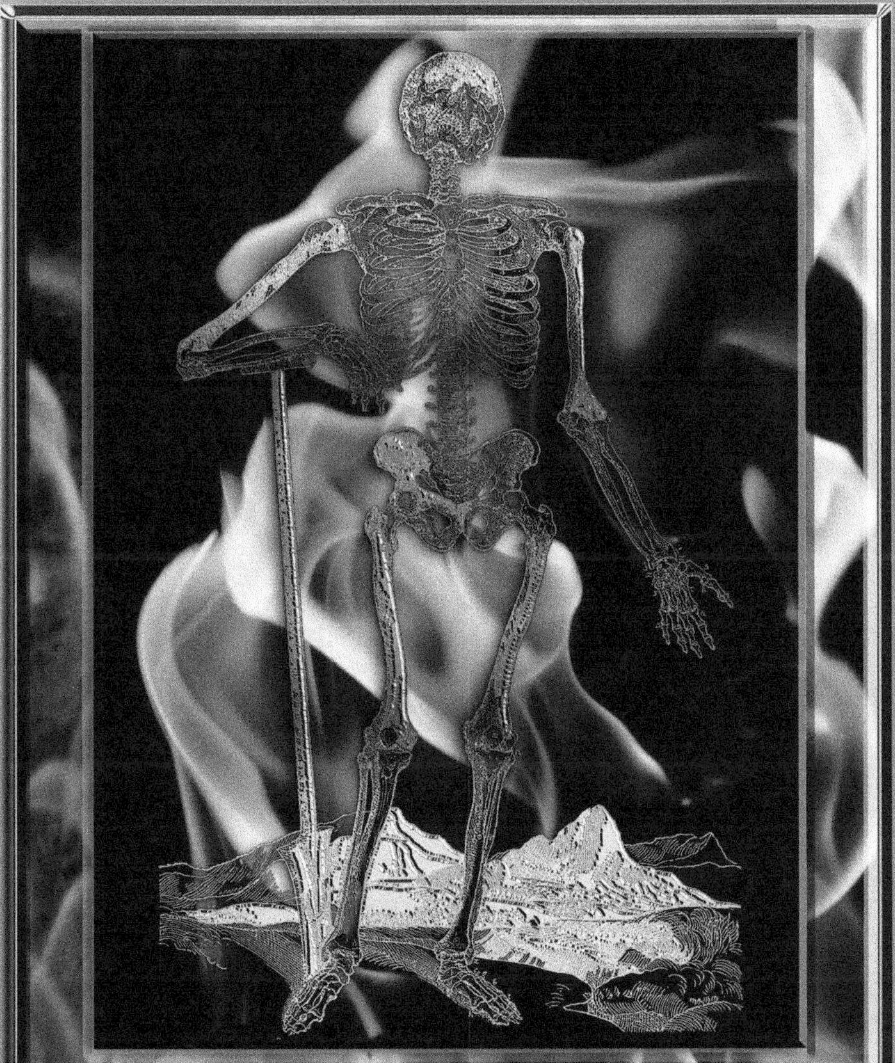

Illuminated Revelations

A few of the now billions of depressed Earthling souls
Had enough energy left to mumble a few words
And so he was able to glean from them
The latest happenings on Earth.

Illuminated Revelations

In 2012, the predicted exponential surge
Of melting ice from global warming
Had quickly inundated all of the coastal cities,
Many of them large centers
Of population and commerce.

Everyone who could possibly make it
Had to retreat inland,
Creating the largest mass exodus in history.

As the heat rose to unbearable levels,
Many had begun living in their basements
As the Earth's infrastructure
Began its eventual collapse.

Millions eventually headed north
Towards Canada and Siberia,
But had to retreat when the ice caps totally melted
And formed the great Ocean of the North;
Most did not make it.

No one but the ignored physicist mathematicians
Had predicted that the end
Could come into sight so quickly.

Then came the dreaded polar shift
That made global warming seem but a small note
Compared to this new and Darker Symphony.

The Earth was thrashed with storms
The likes of which it had never seen;
Electricity was completely out all over the world,
But for a few nuclear powered areas that didn't last.

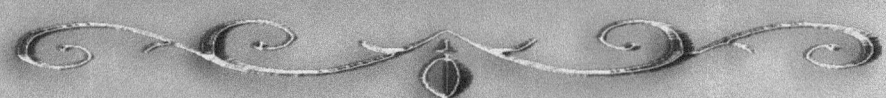

Illuminated Revelations

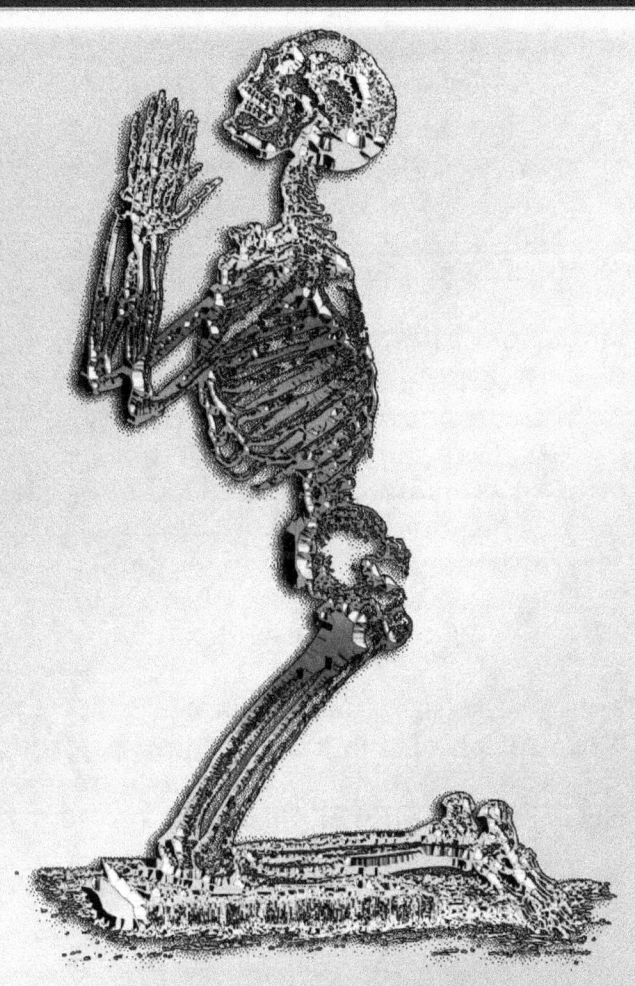

The Great Extinguisher

Our planet is very good at promoting life,
But it is much better at extinguishing it.
Of the billions upon billions of organic things,
99.99% are no longer around here living.

No one could drive very far,
Even on their last tank of gas,
For the roads had melted,
Along with the tires of the vehicles,
And, if the vehicles stopped,
They'd find themselves mired
In the meltdown of the asphalt.

Food would no longer grow very well,
Even in once lush gardens,
In the amounts that were needed,
And, as the heat rose further,
Into the 140s, plant growth ceased altogether,
Although a new but rare
And expensive form of food pill
Extended life for some of the rich
For a short while.

Charon, had, of course,
Seen much of this kind of thing before
From the many other solar systems
And galaxies on which life had formed;
But Earthlings seemed to have
A special charm and hope
Above and beyond the other alien races;
So he rowed and ferried
And deposited them on the far shore,
His job and life forever continuing
In a place with no color,
No joy, and no future—
On the shore of the land
On the edge of oblivion.

Illuminated Revelations

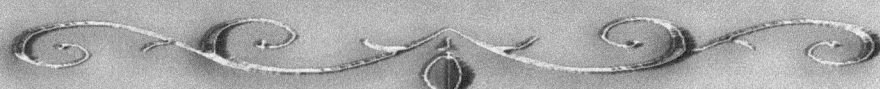

Charon had depths of compassion,
But many passengers might
Many thought him stoic,
Although they were mostly
Beyond the capability.

A sign on the opposite shore said:

Abandon Hope All Ye Who Enter Here

Billions more arrived
In the gray land all too soon
And Charon learned that
Either madness or desperation on Earth
Had caused a nuclear winter all over the planet,
Bringing on a deep freeze that few could escape.

Perhaps they were trying
To combat the ultimate heat,
Which would have been
But a cool breeze in Hell.

The polar shift had greatly
Added to the deep freeze.

...

A few of Charon's still speaking
But chilled customers
Even expressed a longing
For the legendary warmth of Hades.

Charon, stalwart and reliable,
Rowed on steadily,
Ever steeling himself to the misery.

Illuminated Revelations

Finally the masses slowed and dwindled
To a few dribs and drabs over a few years
And then there was no one for several years.

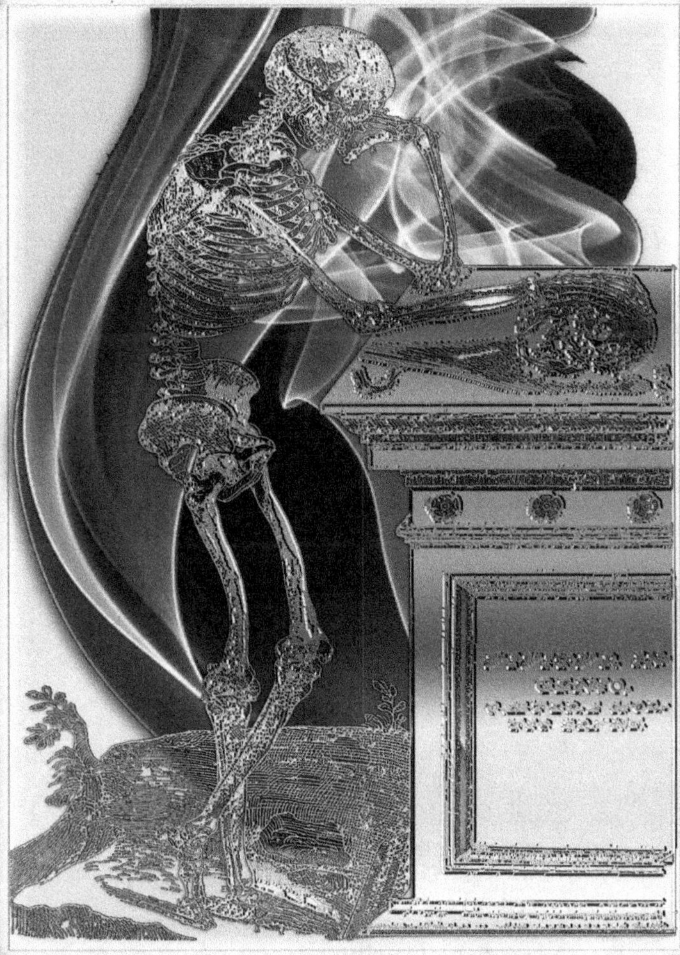

Illuminated Revelations

A lone man appeared on the shore near the ferry dock
And Charon readily approached the man,
Something he had never done before.

They had a long and hearty talk,
For the man was animated
And not at all like any of
The other wretched souls.

"How is it," inquired Charon,
"That you are full of life
And seem to be a good man
But have been sent here?"

"I am not a bad person in any way," the man replied.
"Actually, I just spent some time in Heaven.
I found out there that my sweetheart
Was sent here to you,
For she was a suicide
And so was destined here;
However, I had promised
To be with her forever,
So I chose this place
Over Heaven out of
My love for her."

"Extraordinary," exclaimed Charon.
"I knew the Earth had
A few good men and women—
I've not seen very many clues
Of that elsewhere in the universe.
Did you colonize space—
Will your species continue and flourish
After your Earth bids farewell?"

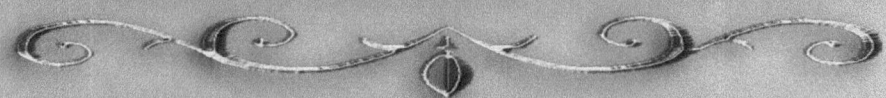

Illuminated Revelations

"I'm afraid not," replied the man,
For too many needless wars intervened
And this greatly delayed our space program."

"A shame," said Charon,
But is there any hope left on Earth,
I mean, are there any others still about?"

"I am the last," the man answered slowly.

The first tear of Charon's long life
Rolled down his cheek;
Nothing had ever made him cry before:
Nothing had ever made him weep.

(Rewritten from Lord Dunsany's brief sketch)

The Love Life of the Glowworm

Flashing desire, the glowfly twinkled across
The starry summer sky, love's energy unspent,
Searching through the darkness,
With passion's might,
for the beacon of her consent—the mating call
Of pulsing, green and yellow light.

At last, came the reply:
Yes, oh yes, a-light, she said;
Now he became a firefly,
As at once she did too.

To a closing flower they together therein flew,
Blinking, winking in the seclusion of its petal bed.
This dance of light and love—their honeymoon
Brightened the night, till it looked much like noon.

Those jolts and bolts, surging, merged in currents,
And swept back and forth as they signaled delight—
fires fuming and oft resuming the flames of love
With electric hugs,

for they had, by now,
Become lightning bugs.

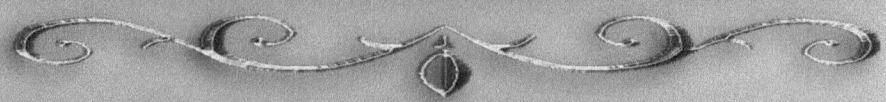

Illuminated Revelations

Illuminated Revelations

ASTRONOMY TOOK THE PLACE OF ASTROLOGY;
CHEMISTRY REPLACED ALCHEMY;
PHILOSOPHY BEGINS WHERE RELIGION ENDS.

Answers

Science discovers the truth everywhere;

Philosophers just sit around in chairs;

Religion just makes for bigger questions;

Evolution explains how we got somewheres.

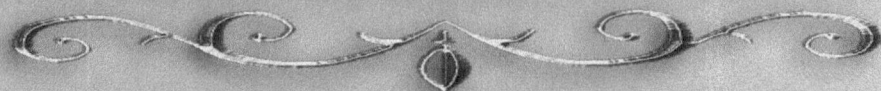

Illuminated Revelations

Natural History

The
Commandments of Evolution
Are unmistakably
Engraved in stone
for everyone to see.
There are no "if's", "and's",
or "but's" in these tablets,
For we can date the rocks of ages.

Ever back through the Ages went I,
Dating rocks and old fossils, by and by,
And found this tablet stone, the Covenant
Of the one and only engraven Commandment.

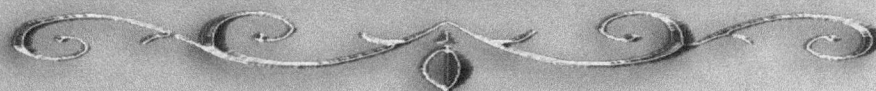

Illuminated Revelations

Readings from St. Austino's Bible 2:1-14

Tip Your Glass, But Don't Spill

The light of Heav'n did the earth illumine,
When God shaped human-nature's acumen.
Temptations He then placed everywhere,
But He'll punish us for being human!

Illuminated Revelations

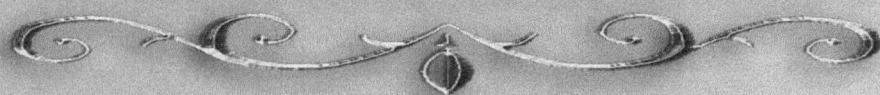

Isaac: Revelations

There's a mote in space known as the Earth,
A pale blue dot of fluff orbiting a hearth;
Due but to Newton's laws of motion, there's
No Godly hand guiding it safe around the sun.

Illuminated Revelations

The Karma of the Barking Dogma

Some Hindus, Buddhists, Christians and Jews
Wondered what stories they should choose;
Even thought they'd already so many chosen,
They just didn't want to keep notions so frozen;

So they met to merge the postulations into one,
Thinking that this might be a whole lot of fun.

"In our hypothesis, there is just the Only One."
"Well, our conception is a multitude of many Some."
"Well, we'll partway meet: there's only the Holy One"
"Nah, the odds of that are over three million to one!"

"Buddha of us was one, so of Gods there are none;
A human above all that now's never seen by the sun!"
"Humph! Holy Jesus of our one God was His son!
He lit mankind's darkness with light of the Sun!"

"No, Jewish Jesus was not of any nature Divine,
But was just a mere man much ahead of his time.
This you all should know, being there at the time.
Look at our history singing those biblical rhymes."

"All is not real, so what is this great big fuss?
Retreat back to where it's all at to slow the rush."
"Oh God's universe and creatures are so real
And that is why we're making this very big deal."

"In the afterlife, we in Hell or Heaven reside."
"Not so fast, for in between these realms we lie,
And if you in this testing life don't do so well,
You'll have so many sub-human tales to tell."

Reason arrived: "Possibility reigned way then back
Before'; there's nothing even holy about all that.
'Tis all made up, those many fabrications made,
So just let it all be, for this is what existence bade.

The Eternally Dead

Here lie the Gods, once so high,
Beneath an electromagnetic sky.

Lo!—the eternally marbled monuments
Of the Moon God, the Sun God (Apollo),
Baal, Zeus, Wotan, Aphrodite, Thor,
Mithras, Isis, Amon, Poseidon, Krishna,
The Druid Gods, and so many more.

Behold!—the ledger of those many Mythologies
That preceded, paraded, and then passed on.

Here they rest, the dead and long gone rhyme,
Adorned with the splendor of mouldering time.

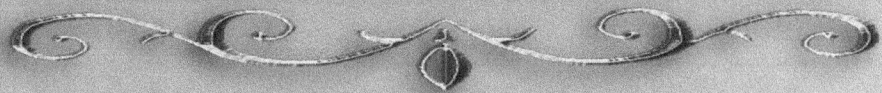

Illuminated Revelations

The Elemental Genesis of it All

If the crucial basis came from somewhere,
Nothing is the only possible lair,
for there are no 'wheres' at all remaining
Of the causeless to be its sustaining.

If the basis is eternal—must be—
Regresses can't go eternally—
Then it is composed of this new Nothing
Because Nothing can have no beginning.

Now, it is not that Nothing ever was,
for it cannot be and it never does,
But that nonexistence ever fluctuates,
Both into and out of its many states.

Of some of these waverings of its form
The opposite particle pairs are born,
for this is all of the natural norm,
As must be concluded by AustinTorn.

!
!
!
v

Swift as existence hastening to its task
Of positive/negative, substance sprang forth
Rejoicing in its splendour, and the mask
Of darkness fell from the awakened 'verse.

On Earth the crystal of the mountain snows
Melted above crimson clouds, and with the glows
Of flame, the Ocean's horizon arose,
With flowers in fields or forests which unclose

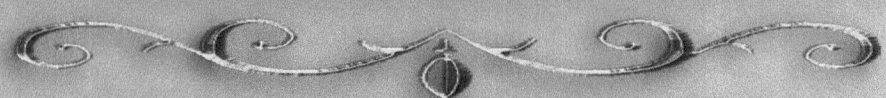

Their growing vision to the kiss of day,
Swinging their censers in the element—
Of eastern incense lit by the new ray—
Burned slow and inconsummably, and sent

Their odorous scents up to the willing air;
And, in succession due, did continent,
Island, sea, and all things that in them wear
The face and complexion of mortal flair,

Rise as the sun their father rose, of old,
In portions of the soil, which he did mold,
As his own, and then imposed, yet untold:
All their thoughts that must ever now unfold.

THE PYRAMID

The Electric and the Magnetic forces
Transition into each other,
As a self-renewing electromagnetic wave
That can go on at light speed toward 'forever'.

These points are like as East and West
Blending into each other on the globe.

The opposing forces of Weak and Strong
Are of changeability vs. stability—
They are as different as North and South

Gravity arises from all the forces together.

The Strong and Weak opposition
Perhaps shows the separation of the forces
Once thought to be able to be unified.

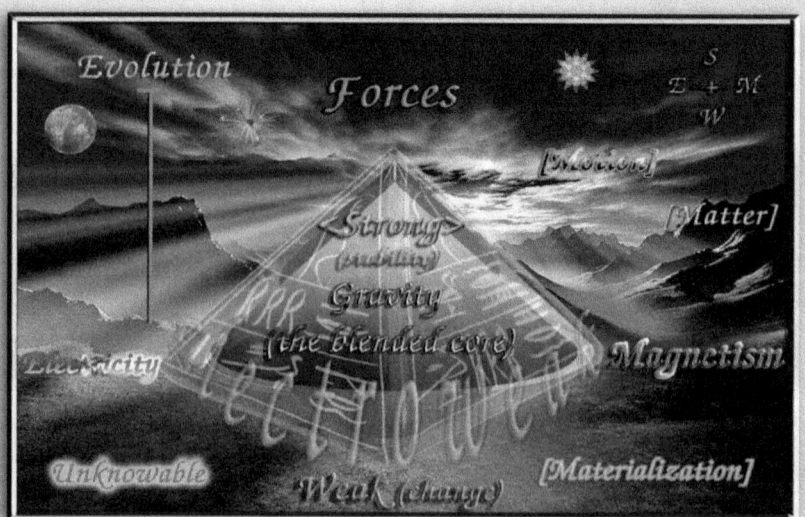

WHAT REMAINS OF MAN'S WORKS? TIME AND SAND DEVOUR THEM ALL.

MAGICAL HAPPENINGS

What secrets of life and death
Lay buried in the sands?

What inaccessible truths
Protect themselves by their own magic?

The old Rascal lit up a cigar,
And the stories unfolded
In the haze of this pipe dream…

"Do tell what else
Was in that Great Pyramid, Fredrick,"
The General suggested.

Illuminated Revelations

"There were 4000 year-old iron weapons
That did not rust,
Looking as new as
The day they were forged.

"I held glass that bent without breaking.

"I drank from a vase
That poured water without end;
I filled an entire tub from it
And bathed away all my dirt and dust.

"A compass needle went around
And never stopped.

"I ate a cake but I still had it.

"I saw the starry skies
Through solid rock walls.

"I entered a room that had no door.

"There was light within the room
But no flame or openings.

"I looked into a grain of sand
And saw eternity."

Fredrick paused, recalling.

"Outside, I saw the Sphinx.
Its glance was fixed on something else.
It was the glance of a being
Who thinks in centuries and millenniums.

I did not exist and could not exist for it,
For it was the face of eternity."

No one spoke.

The General rose.
"Next, after an hour break,
During which you might go out
To see the scenery,
We may hear some about a long trek
From an escape from a Soviet prison
Through the mountains
And across some ink-black rivers."

Questor and Top Secret
Headed down one
Of the many paths of Niihau,
Its secrets ever shrouded in mist from above
And all around from the other islands;
But, here they were,
Within the Forbidden Paradise...

"Wow!"

"Come back, friends,"
Said the General,
"To hear of the dark,
The light, and the never."

"We are here, being ever."

Illuminated Revelations

(Bronto at the shore)

(Niihau path)

"There are books unwritten and never told."

"We can listen until we get old."

*"By what muted shore of the dark river
Did its strand call me forth?"*

"We're sure that we'll never hear worse."

*"By what far edge of furrowed forest
Didst the Motherland seek my name?"*

"Oh, Dragon, through what hazy depth
Of gloom hast thou tread and threadest?"

"Gather thee round and you shall knowest."

Illuminated Revelations

PATENTING THE TAO

All is solved with the TAO,
With just a few word-details now
For the lawyers to work on,
Such as with the TAO
That is not the OAT.

*I tried to get a copyright
On the TAO symbol,
I guess God owns it.*

We will still get a PatPending, Prof—
It's in the works...

Note:
The lawyers don't like to say
That the TAO is not a duck,
The TAO is not a dog, etc.,
On up through everything
Real and imaginable.

We will also have a logo of sorts,
Maybe a cube,
With some letters on each face
To show that we've covered everything,
Like, for one face:

G U T
D O A
N E O

Illuminated Revelations

(The Path That Cannot Be Spoken)

Illuminated Revelations

Taoist propriety and ethics emphasize
The Three Jewels of the Tao:

Compassion, moderation, and humility,
While Taoist thought generally focuses
On nature, the relationship between humanity,
And the cosmos, health and longevity,
And wu wei (action through inaction),
Which is thought to produce harmony
With the universe. —Wiki

(God is not really associated with the TAO)

OK, Prof—I was up late,
Burning some of your midnight lamp oil
At the Supreme Court.

The Good News:
God did not show up to claim the TAO rights.

The Bad News:
The court's lawyer is named Fredrick;
He's the attorney general for the universe.

The Not So Bad News:
I am the PatTorney general,
One letter higher than Fredrick,
Making him a peon.

They said that if the TAO
Couldn't be spoken
Then I didn't know
What I was talking about.

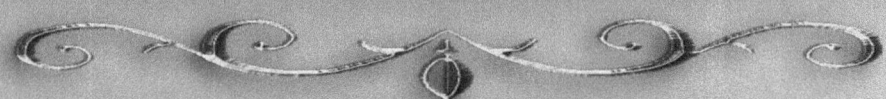

(The Courthouse)

So, I presented the +/- balance
Of Nature's Account
And they liked it.

Some excerpts from my presentation:

From a Western perspective,
The Taoist view of sexuality
Is considerably more at ease.
The body is not viewed as
A dangerous source of evil temptation,
But rather as a positive asset.

Illuminated Revelations

Taoism rejects Western mind-body dualism;
Mind and body are not set in contrast
Or opposition with each other.
Sex is treated as a vital component
To romantic love. — Wiki

(What do the acyronyms stand for?)

Illuminated Revelations

Illuminated Revelations

Can you resist the beauty of loves truth
When roses and tulips bloom in loving hearts?

Illuminated Revelations

Love

{The Greatest Day on Earth}

But, this is to be the day of the new moon;
At least there is a chance, thinks Percevale.
They arrive on the shore of Iceland,
And, on this day, as on every day for a month
Either way in this northern land,
The sun does not rise,
for it did not set the day before.

Just before noon, strange bands of shadows
Begin to rapidly cross the land
And Percevale feels that perhaps the end is near.
The ground begins to shake and heave
for a few moments and then all is silent,
So very silent as to strike one dumb.

Something terrible seems to be happening.
Grazing animals look for shade trees
And lie down to sleep.

Then, about noontime,
The shadow of darkest night covers the land
As the moon begins to kiss the sun and cover it—
It is a solar eclipse!

Merlyn's old notes in the archive were accurate!
Thank the gods for the old wizard!

Illuminated Revelations

Wo[man]

Illuminated Revelations

During the seven minutes of total darkness,
Percevale sees a black disk in the sky,
Surrounded by faint wisps of flame.

It is, of course the new moon in all her black glory;
Indeed, the new moon can only be seen
During a solar eclipse, and never at any other time.

And there near the sun is a bright 'star'.
It can only be the planet Mercury!
Yes, there it is, in plain sight, at high noon.
And farther out, Venus can be seen!

Now the ground begins to really shake, and
Percevale hurries to his ship with the Ice Maiden.

They leave Iceland but see the volcano erupt;
Rocks are flowing to the sea like water!
But, the water puts out the fiery flow
And so they do not see fire in water,
But just a lot of steam.

Then a tremendous plume
Of smoke and debris is sent up into the sky
And is carried south by the unusual winds
Born of the marriage of summer warmth
And ice cold air brought on by the blockage
Of the sun's rays by the dense volcanic ash.

The spontaneous cold front sweeps south to Gaul

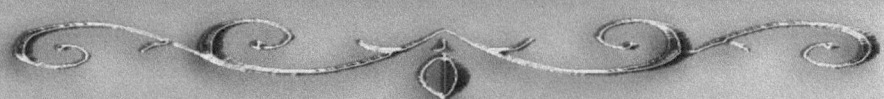

Illuminated Revelations

Illuminated Revelations

On the reversed upper winds,
Bringing the darkness of the ashen sky with it.
As no sunlight can penetrate,
The air below grows colder and colder,
And what would have been rain now turns to snow
Over Cisalpine Gaul for a brief time
Before the westerly winds can disperse
The volcanic cloud around the earth.

That evening the sun sinks low, but does not set.
On the water is the glitter path of that fiery ball—
And so we have fire in water!

The sun has kissed the moon,
And Percevale gathers the Ice Maiden
Into his arms and kisses her,
His capacity for love far from dead,
But growing stronger every minute
Of this glorious day
As both of their curses fall by the wayside.

(Taken from the Celtic Chronicles,
found in an iron box beneath an Abbey.)

Opportune

Be wide aware when chance shines as your sun,
for she, in turn, happens on everyone—
Graciously welcome the lady of luck
By recognizing her as Dame Fortune.

The Root of All Evil

Other than direct hurts to persons,
Is what some groups think of 'good' arbitrary?
And harmless, until it is imposed on people?

We see many good and bad things directly,
Person to person,
Via the actual.
Such are the good civil laws
And good human values taught.

The problem becomes when we 'see'
From no direction but the imagined,
Via the unreal.
These 'good' things, merely pronounced,
Also define their 'bad' counterparts.

One then 'forgets' their null source,
Leaping into complete adoption,
Becoming more and more with them one;
Thus, the ideas must be protected.

Anger arises toward the contrary,
As emotion stains the brain.

Then, evil is done
In the name of 'good'.

All these 'good' things
Eventually
Come to a bloody end.

The Roots of Evil

Nip trouble in the bud, lest it grow
And sprout like a weed, blossoming with woe,
And spreading, thickening all around, till
It imprisons you like some old hedgerow.

Problems are not as complex as you think:
Simply, misery and death follow drink;
Evil and cruelty are the same reflected;
Drugs plainly lead to a life out of sync.

Illuminated Revelations

I have recently come across
Some new revelations
And have derived more of the secret...

HOW THE ALLIES WON WORLD WAR II

Warner Heisenberg, the head of
The German Nuclear Weapons Effort,
Was full of the uncertainty
That he had discovered in physics.

Heisenberg was entangled with his old mentor,
The Danish physicist Neils Bohr,
They being old friends, like father and son.

They were also supposed to be enemies,
For Germany occupied Denmark.

Together they had created a physics
Of deep truth and beauty,
For beauty was the expression of truth.

They also made possible the physics
To destroy large cities, even the entire world.

In 1941, Heisenberg went to see Bohr,
The 'father of quantum mechanics',
In Copenhagen, Denmark;
But we don't know what they discussed;
Yet, Germany failed to complete its work
To build an atomic bomb.

Did Heisenberg deliberately withhold
Information from the Nazis?

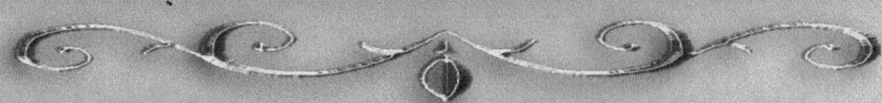

Did this consummate mathematician
Neglect to perform
An obvious calculation?

Did he, with Bohr,
Form a complimentery pair,
Joining their views
Of the political position
Versus its velocity
To form a complete picture of reality?

Did a man's heart turn the tide of War?

(So we shall see...)

(Correspondence and Complemtarity)

THE DRAWING

On September 9th, 1943, Neils Bohr
Walked to a meeting place near the water
And crawled in complete darkness to a beach,
For the gestapo in Copenhagen
Were about to arrest him.

He secretly crossed the Oresund to Sweden
And remained there until October 6th,
Wherefrom the British flew him to Scotland.

That evening,
Sir John Anderson gave Bohr
A briefing on just how far
The Anglo-American atomic bomb
Program had progressed.

Fermi's reactor had begun operating
On December 2, 1942.

Bohr was shocked,
For he knew that only
The very rare isotope uranium 235
Had fissioned in the
German Hahn-Strassman experiments.

This was fully two years after Bohr
Had met with Heisenberg
In occupied Denmark.

What had the Germans done during this time?

No wonder Bohr was alarmed.

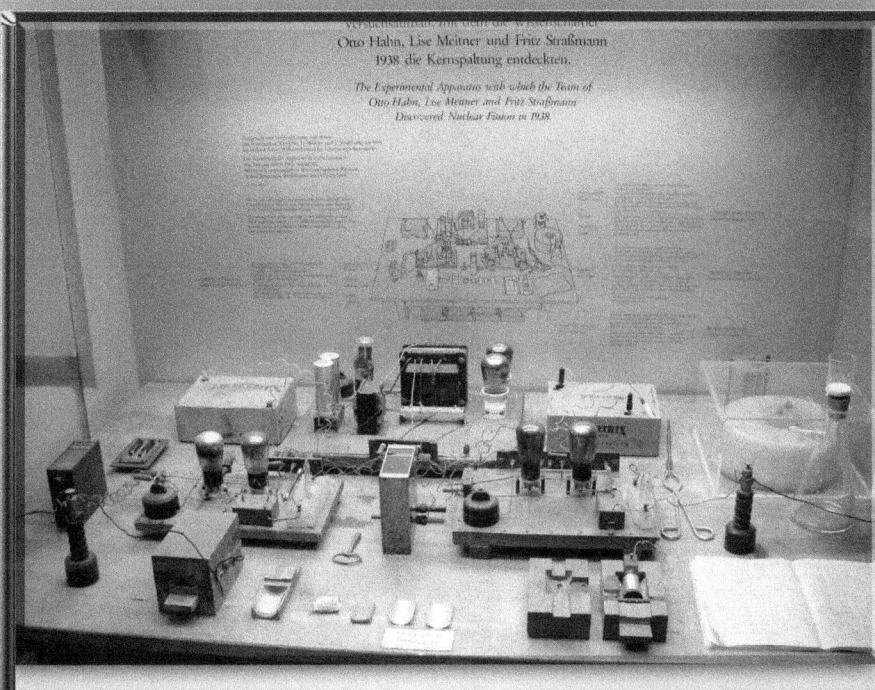

And, yet, Bohr, somehow, had a drawing
Of the German nuclear reactor,
Which at first he thought
Might be the weapon itself.

All knew that plutonium,
Which does not exist naturally,
Could be chemically separated
From its uranium matrix
After bombarding a reactor's
Uranium fuel rods with neutrons.

The critical mass was not in tons
But in pounds, prompting the Allied effort,
Not so much Einstein's letter to Roosevelt.

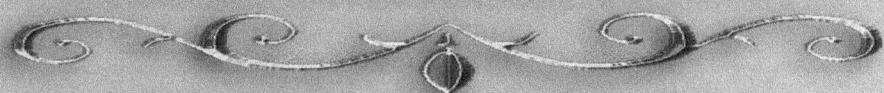

Illuminated Revelations

Bohr went to work at Los Alamos
Where Oppenheimer was orchestrating
The impossible from 1943-1945.

On New Year's eve of 1943,
Scientists looked at Bohr's drawing
Of Heisenberg's nuclear reactor
In Oppenheimer's office.

Within two days, General Groves,
The military commander of the project,
Received a document beginning with

"The proposed pile [reactor]
Consists of uranium sheets
Immersed into heavy water."

And ended with

"The arrangement [the drawing]
Suggested to you by Bohr
Would be a quite useless military weapon."

By late 1943, nearly everyone in
The German nuclear program,
With the exception of Heisenberg,
Had become convinced that
Uranium plates were inferior
To a design using rods or cubes,
For the most efficient design
Involves separated lumps of uranium
Embedded in a lattice within the "moderator";
But the worst possible solution
Is placing uranium in sheets or layers.

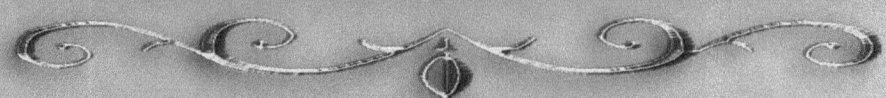

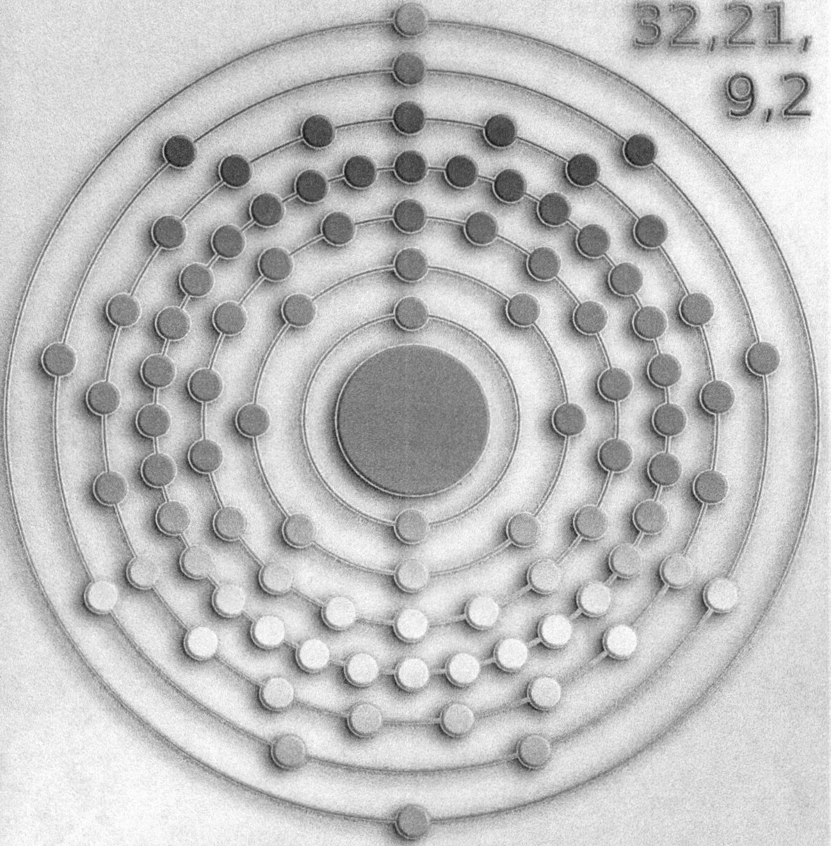

Illuminated Revelations

92: Uranium

2,8,18, 32,21, 9,2

The role of the "moderator"
Is to slow down the fissioned neutrons,
With only heavy water or carbon
Seemingly being feasible.

The Germans had chosen heavy water,
Its separation from ordinary water

An expensive and difficult process,
Since carbon graphite is
Rendered useless by an impurity
Of as little as one part boron in 500,000.

At Los Alamos, Leo Szilard
Was a fanatic about
The purity of the graphite,
And, since it was readily available
They decided to use it for carbon.

The Dragon's breath was unleashed.

Replica of The Fat Man Atomic Bomb at
the U.S. Air Force National Museum.
Photograph credit: www.ngmalalpha.com

("Fat Man")

None of the German reactors
Ever even operated.

Where did Bohr's drawing come from,
For it had "Made in Germany"
Written all over it?

...

It could have only come from Heisenberg.

(Bohr and Heisenberg)

THE FURTHER WHIMS OF FATE

In 1935, Fermi had almost discovered fission
Three years earlier than Hahn-Strassman;
But, in order to shield the detectors
From unwanted radiation
From the slow-neutron process
He had covered the uranium target
With aluminum foil.

This prevented Him
From seeing the very energetic pulses
From the uranium fission that was taking place.

Thus, the race to build an atomic bomb
Might well have started in 1935 rather than 1939.

If so, World War II
Could have been nuclear from the beginning
Or even have become a Cold War—
All of this not happening
Because of some aluminum foil!

Illuminated Revelations

(The Enola Gay)

(The Solvay Conference, 1927)

THIS TOTTERING EXISTENCE

So called "empty" space is vital,
For that's where there's the recital
That forms and plays the tunes of reality,
The grand cosmic symphony—
As existence fluctuates with the non,
Those causeless waverings of undulation.

It was once thought that the shove
Of this total energy was of
The order of 10**120 orders of
Magnitude above.

Well, if that were so near,
Then we couldn't even be here;

It was the worst calculation
In all of scientification;
So, we weighed the universe,
Summing all of its constituent verses.

The universe weighs nothing at all!

This, too, since we found that
Our universal space was B flat—
Not just via the 60 degree angles
Of a very small triangle.
Not even using stars,
One that went from here to Mars
To Venus and back,
But all the way back
To a degree of the CMBR,

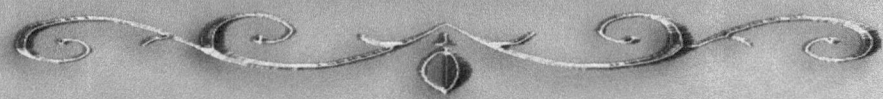

Which represented 100,000 light years,
And measured the curvature:
The rays didn't converge or diverge.

The ultimate of this geometry
Is that being flat is a beautiful symmetry
That leads to yet another beauty: zero.
The ever returning, conquering hero.

Far from being the *Magnificat,*
We are more insignificant
Than we ever imagined, even Kant,
As all is a big nothing,
But also, since, considering
That all the specs of matter's amount,
For whatever is the measly count,
Compared to dark matter and dark energy
Are but a kind of pollution, irrelevant, really.

THE SIMPLE BASIS OF BEING

As for forces, which is just a prelude here,
We note that two of them are transitional,
The Electric and the Magnetic,
Each giving rise to the other;

And that two others are oppositional,
The Weak and the Strong,
The Weak promoting changeability,
The Strong promoting stability.

Gravity is then left as a blend of all.

(Strong vs. Weak) [Gravity] (Electro <—> Magnetic)

So, would oppositional and transitional pairs
Work for our human being as well?

For us humans, all is of the *Movement* of *Appearances,*
These *Movements* giving rise to notions of time...

(Past into Future,
Or the Then to When through the Now),
Transitional in only one direction;

While *Appearances* beget notions of
Matter lumps, in a place of Space...

(Matter and Space, or the What and Where),
A kind of an opposition in that
The knots of matter are separate
From the gaps of space in between;

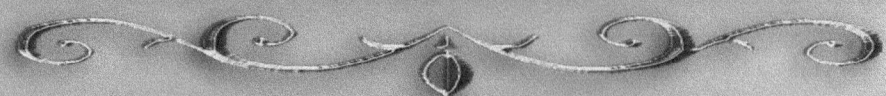

Or, in short, all seems to be the
Movement through time/distance
Of Matter in Space.

(Matter vs. Space) [Being] (Past —> Future)

nothing (why) + possibilities(how)

{ [space(where) <— (appearances) —> matter(what)]

+

[past(then) —> (movement) —> future(when)] }

=

the spirit of life

evolution

being(who)

now

We will see that our being is composed
From these simple notions begun,

For *movement* grants time—
The Then and the When
Of the Past and the Future,
Via change;

While Matter is the What,
And Space is the Where,
Via 'clumps'.

The blend of all these would be
A kind of spirit of life.

These fields then further combine:
The What/Matter + When/Future field
Becoming the Progression
Of matter into the future,

And the What/Matter + Then/Past field
Being the History
Of the matter past—what has occurred.
The When/Future of Where/Space field
Makes for Wishes, hopes and dreams
In the future place of space;

While the Then/Past + Where/Space field—
Begets Remembrance of memories
In the past space.

The emergent fields then further combine:

Learning becomes of <u>Remembrance</u> and <u>History</u>;
A Change of Outlook becomes
Of <u>Remembrance</u> and <u>Wishes</u>;
A Change in Structure is <u>Progress</u> from <u>History</u>;
And *Vision* is of <u>Wishes</u> and <u>Progress</u>.

Then at the next higher stage,
Being **Creative** is brought forth
From *Learning* combined with a *Change in Structure*;
Direction results from *Learning* and a *Change of Outlook*;
Growth is the *Vision* for a *Change of Outlook*;
Planning is the *Vision* for a *Change in Structure*.

Finally, **Creating**, **Direction**, **Growth**, and **Planning**
Compose one's being—The Who.

Illuminated Revelations

TO THE DEEP

To learn the Secrets—what IS and ev'r WAS,
We must brave the crypt and ghost of cause...

So, into the deep, we go, without pause,
To look down, ever down, no self to keep—
Through birth, death, and the shade of sleep,
Through paths unkempt, underswept—to the deep,

Through the cloudy strife
Of this hazy life,

Through the equations of eternity,
Their non-paternity nor maternity,

Past the realm of the things which seem or are,
Even o'er the steps of the remotest bar.

Down, down!
Where the mind whirls round and round,
As the ear draws forth the sound,

As the eye sees the light,
And of the dark the fright.

Down, down,
Beyond all death, despair, love, and sorrow,
Past yesterday, today, and tomorrow,
The body's guide is but the logic of the mind.

Illuminated Revelations

Illuminated Revelations

Down through the fog, the not, and the void,
Where "God" and Nothing fail; Oh, zoids!

Down! Where reigns the night, and the air is thin,
To where the sky and stars are not, but within,

Where the radiant have not their throne,
Where there are one or some pervading, all alone.

Down, down! To the fathoms of the cryptic;
Down, down! Where substance slept with arithmetic,

Toward the spark yet nursed by embers,
To the first and last that the universe remembers,

To seek the gem that shines—the wealth of mines,
The jewels so treasured by thee and thine.

What accelerated life's momentous gem,
Letting your motto be "Carpe diem"?

What seized the moment or lost its momentum?
Wearing not the time as its royal diadem?

World does not pass by—we pass through it;
Clear your being so the treasure may arrive;

This spirit sparkles of a different light—
The gemstones are of a different mine.
Down, down!

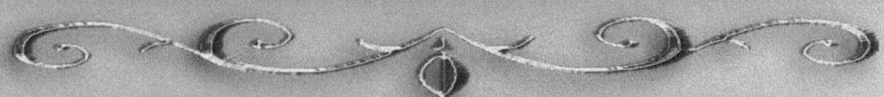

Illuminated Revelations

Carpe Diem

We guide thee, we must carry thee;
Down, down!
We're illumination beside thee…

Fear not the proof—it's the beauty of truth:

What entropic seas never-endingly
Cast Deathly Time aside,
Ceaselessly thriving on…

Of that which was the imperishable—
The flame of beauty inextinguishable,
Forever celebrated as immutable…

That gained its perpetual permanence
From the undying love of the glorious truth?

As, once, above the ground, you're born again
When the roseate hearts were cleansed by dew;

And lucky were you if spring found you new,
As every blossom on the bush blew full;

When these wonders the new morning bestrew;
The beauty of truth was all that you "knew"…

For, life's hardships were softened by beauty,
All its weaknesses strengthened by truth.

As when roses blossomed, like realizations,
Beauty itself bloomed from the well of truth.

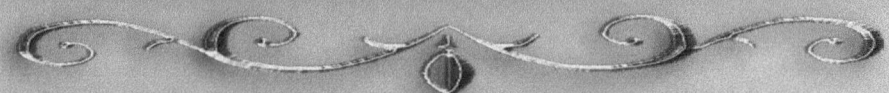

Illuminated Revelations

When a deep truth is known so intensely
That all of its clothing falls away,
Then we have learned the beauty of truth, for
The reality of meaning is beauty.
Life, although anguishing, must be lived fully,
Since, if we're alive enough to feel its beauty,
Then we're exposed to the opposite twin—
Yes, Beauty's other side is Melancholy.

Illuminated Revelations

Down, down!
For now, rarely enough, existence is left aside,
And, yet, the essence has its other side—

Life, although anguishing, must be lived fully,
Since, if we're alive enough to feel its beauty...

Then we're exposed to the opposite twin—
Yes, Beauty's other side is Melancholy.
Down, down, the essence beckons us home,
As like the contained-container is the poem.

When a deep truth is known so intensely
That all of its clothing falls away...

Then we have learned the beauty of truth,
For the reality of meaning is beauty.

Opposite twins rule the causing call,
The positive and negatives being all.

When sadness brooded over the morrow,
One visited the deep well of sorrow.

There enshrined, inseparate, Beauty said,
"'Twas from me that sadness you borrowed."

Do we live the life of art,
Each playing our part?

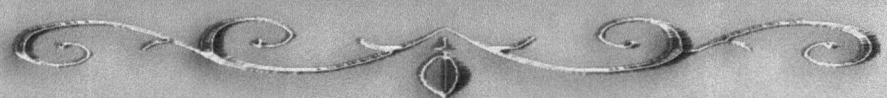

Illuminated Revelations

Illuminated Revelations

Nay, that is not life, nor a part, bit,
For there's another dimension to it.

Art and poetry enrich human experience,
But they're no substitutes for the living of it.

Like Keat's figures on the urn, should we live life less?
NO!—because what is deathless is also lifeless!

Down, down!
Truth and beauty must be inseparable,
Although seemingly imponderable.

On that sphere above,
Soft breezes blew, caressing me and you,
As we kissed the roses and drank their dew.

Reason and passion merged into one,
As truth and beauty made their rendezvous.

Down, down,
Through all antiquity, past all of the known
Arriving at the lowest, remotest throne...

But not one of the highest perfection,
For that will become of the opposite direction.

Here, the enigma of the immortal
Is undone and unloosed, through its portal—

...

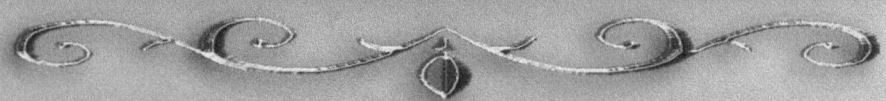

Is your past imperfect?
Is **your** future tense?
Give yourself a present.

Illuminated Revelations

The Theory of Everything mortal—
An Idea for which we've opened the door to.

Down, down, to the end at last!

Here the timeless, lawless, and formless
Of the unordered, uncreated scene;

Here the causeless reigns "supreme".

Illuminated Revelations

The End, the Beginning, and All that Lies Between...

Illuminated Revelations

The Clarity Of Reality

Illuminated Revelations

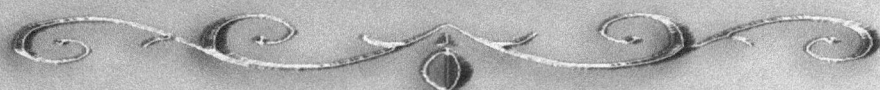

THE CLARITY OF EXISTENCE

This is a short series on the clarity
Of objective language for describing
And explaining existence,
While at the same time heading
More towards really explaining existence
(Which has always been THE QUESTION
And is much different than
Merely describing existence).

What is not really known will not be used
(As in "made up" for much of Big Bang theory,
Which, by the way, explains nothing
Of why there is existence
Or what was before the Big Bang,
How it conserves energy, etc.).

So, no unknown precursors will be employed.
This leaves out much more than God.

Hopefully, some math may confirm the logic,
As is always the best case.

Existence is that which has a **quantity**.
We need not to be able to measure the quantity
To a jillion decimal places,
For it is quite sufficient that the quantity
Can be measured at all.

So, existence is what is real and quantifiable,
Being physical and/or material,
But I will just call it existence;
However, I will always define the quantity.

Who Am I?

I am the balance of nothing,
Come from nowhere,
Residing here in its midst
And middle of finiteness.
Within a parentheses
Of "All That Is"
Of ever-during Eternity
And unbounded Infinity.

We'll start with the expansion of language
For the existence of **space**,
It necessarily being physical,
But not material,
Even though it contains
Material field everywhere—
The 'everywhere' being the 'space' part...

While space is physical but not material
It is quantifiable and therefore it exists—
Its well known quantity is volume.

Space is the most inert medium imaginable—
It does not directly interact with the material
(That material realm even consisting
Of mostly empty space).

Space provides a framework for reality
And so it is not on a par with, say, energy,
As a constituent of of reality itself,
Yet it is still physical,
Although that is not widely accepted;
However, General Relativity requires
Curved space around massive objects
(Einstein even referring to matter
As "curved empty space". Is it?).

Space is just as real as energy
Because volume is a quantity.
Energy density is even defined
As energy per volume—
As they must both exist in par
For this equation.

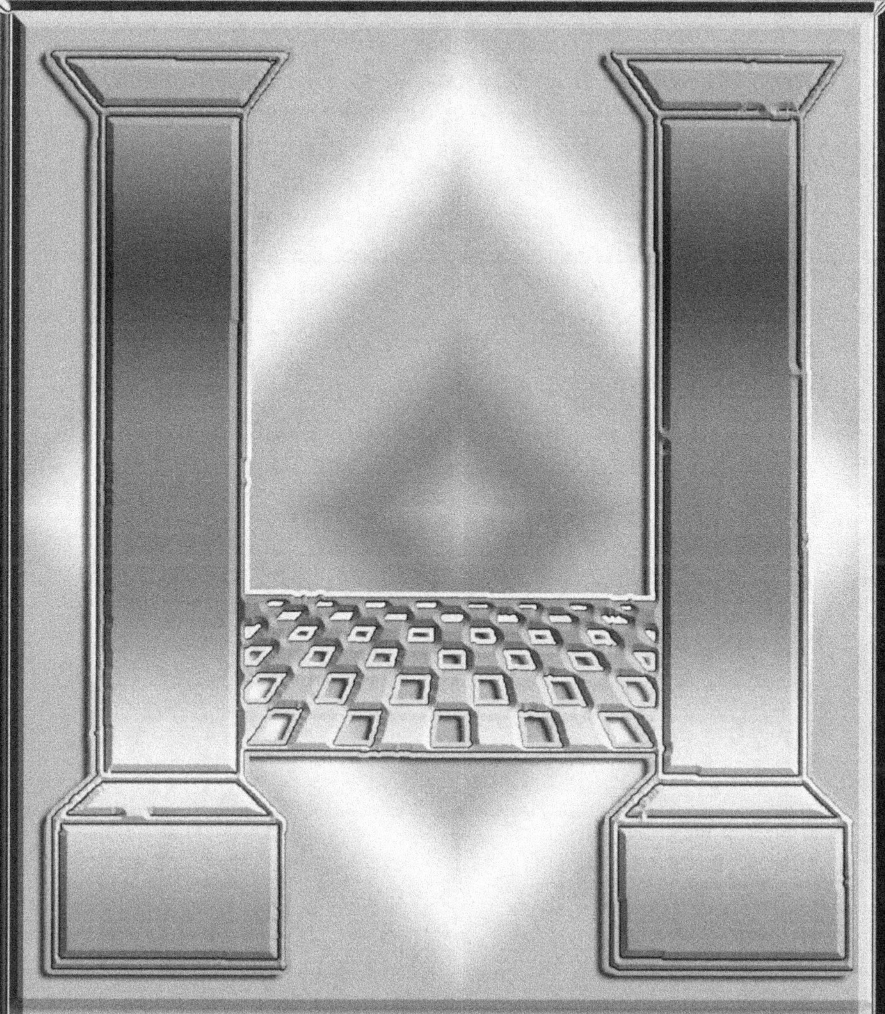

Illuminated Revelations

Space need not have full interaction to exist,
Just as two laser beams crossing do not interact
(A weakness of photon-photon interactions);
So, existence is not circumstantial,
But ever only quantifiable.

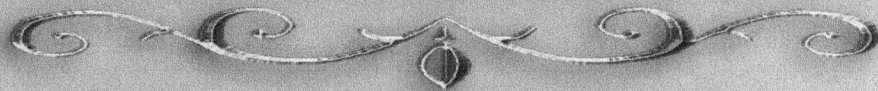

Yet, space is space and energy is energy
And it seems that the twain shall never meet,
Although one could say that
They still interact geometrically
Since the presence of space allows
For the distribution of energy.
Without this "interaction",
Reality would be a single point
Of infinite energy density.

Space must be forever,
Being infinite in all three of its dimensions,
Which is really like Infinity**3, whatever that is.

Illuminated Revelations

Illuminated Revelations

Fields throughout space would then
Be the [new] aether of space, not space itself,
Although each supports the other, hand in glove.

I must resist calling fields 'immaterial',
As some may call them,
For fields are surely of the material spigots: matter,
And, furthermore, inseparable from matter.

As for **forces and pressures**,
They, too, exist, but are secondary,
And so they must be called 'cosmetic',
For they are not directly linked to reality's composition.
Think of this 'cosmetic' as being characteristics
Of energy forms or relationships
Between more substantial quantities.

Space is **conserved** since its volume
Does not vary when[if] it is distorted by matter,
And, at any rate, this volume
Can neither be created nor destroyed.

To know what the universe is made of
Is too look at its conserved quantities,
For everything else is cosmetic.

Conservation is evidently vital
To the universe's configuration.
We will look at this later.

Existence, then,
Is the set of all quantifiable things—
Some material and/or physical,
Or as cosmetic—and relational.

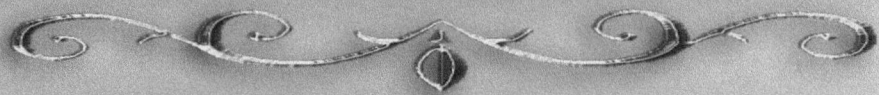

Illuminated Revelations

(No unknown precursors)

The sum total of quantity is reality.

Time exists in terms of seconds,
Energy in terms of joules,
Velocity as a ratio of quantities
(Time and distance).

Nonexistence would be to have no quantity,
This single characteristic being the only difference
Between existence and non.

These definitions are not simplistic,
For the deepest level of reality must be simple.

Illuminated Revelations

The universe is a complex equation,
But the most perfect one possible.

Headings

An incomplete solution
Is invariably an incorrect solution.

I haven't forgotten about the series
On the clarity of all existence;
It's just that it's the slowest going piece
That I've ever done,
Even with much informational help,
For it takes quite a while
To figure out all of reality
And "Why existence",
An understatement
If I ever heard one.

Yet, it has an answer
And that must be the simplest thing,
For there can be no more basis below.

The answer to the equation of the universe
Lies in symmetry, and, of course,
The necessary conservation laws within that,
And, although simple, it will be astounding.

We can't just give answers without showing,
And so that's why we will have note some
Fundamental descriptions of existence,
Although the primary aim is explanation.

This is because what the descriptions
May have in common will
Lead us to the simple explanation
Of the basis of reality's bedrock.

Illuminated Revelations

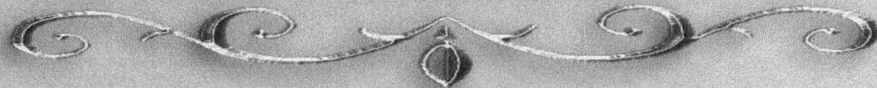

(The Reality of Existence)

Illuminated Revelations

The infinite universe will be seen
To be all there is,
It being eternal in both directions,
Past and future.
(And so we have Poppa's "Infinity for Eternity"!)

Nor will the universe
Be expanding, contracting,
Or even cyclic in the typical sense,
But more like the galaxies recycling
The elements back to hydrogen
Every so many large number of years.

All this, then, we will find,
Due to the symmetry of conservation
Of volume, energy, momentum,
And so forth.

Space will be found to be
Continuous (at higher levels),
We will understand
Why there is energy everywhere,
We will see why
We must exist in the finite region
Between the largest and smallest infinities,
And we will learn why the universe
Can <u>only</u> be the way it is.

We must discard the Big Bang theory
As it not being able to conserve energy
And because there could be no inflation
Or expansion—
Expanding into... what, exactly?
Some region already present by definition?

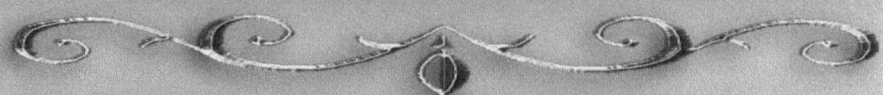

Illuminated Revelations

So, too, will recessional red-shift speeds fall,
Quantum haziness (not undone by observation),
String theory, non steady-state, and
Many of reality's notations other than
The dimensional distances and time.
(Of our 3D+T realm which is exactly as such)

We will see,
Even beyond Emmy Noether's First theorem,
Why symmetry and conservation must be so,
And that
The universe itself
Is the source of all data and knowledge as
The direct reflections of reality's underlying structure.
(And not data and knowledge of human creations)

Illuminated Revelations

The Bird of Time

(What is Time?)

Noether's First Theorem

First proved in 1915 and published in 1918,
Amalie Emmy Noether's First Theorem
Gives a profound connection
Between continuous symmetries
And conservation laws
For certain classes of theories.

The familiar consequences of Noether's Theorem
Are that space translational symmetry
Gives us conservation of momentum,
Rotational symmetry gives us conservation
Of angular momentum,
Time translational symmetry gives us
Conservation of energy, etc.

More carefully,
Noether proved that a physical system
Described by a Lagrangian invariant
With respect to the symmetry transformations
Of a Lie group, has,
In the case of a group with a finite number
Of independent infinitesimal generators,
A conservation law for each such generator.
(These, luckily, are what are used in physics)

If we have the case of
A countably infinite number
Of independent infinitesimal generators,
We still arrive at certain,
Profound, dependencies.

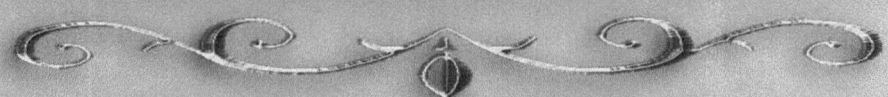

The Equation of Reality

Good-evil, on-off,
hot-cold, man-woman,
Up-down, left-right,
here-there, past-future,
In-out, proton-electron,
time-distance,
A—>B/B—>A, +n/(-n),
action-reaction, (=),
Nothing-Infinity,
something-lack of
Mass-energy,
existence-nonexistence,
Particle-antiparticle,
end-beginning ...
Neither side could exist
without the other.

Unbounded Space:
The Large and the Small

There can be no boundaries to space
Because a boundary is a quantity—
And a quantity, as seen, is existence.

Space is unbounded,
Although we will later see
That it is closed in a way
Much more profound than its size.

The universes's size
Is the largest possible
Quantity of existence
And is fixed throughout time,
Having the same volume today
As it had yesterday
And will tomorrow.
(A necessary symmetry)

So, too, must it be flat
And shapeless.
(More symmetry)

While from the standpoint
Of symmetry in three spacial dimensions
The totality of the universe is cubic,
The universe cannot be the shape of a cube as such
Because it has no boundary…

And the same for an infinitely small "point":
It must have no form since
That is a consequence of infinite smallness.

Illuminated Revelations

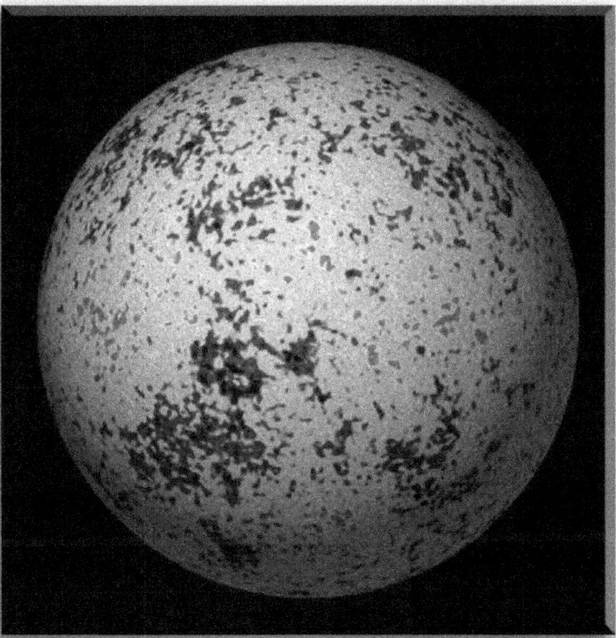

Our Finite Existence

We are suspended here
In our finite realm,
Where we must be—
In a balance of
Infinite largeness
And infinite smallness.

Of course, we have knowledge
Of the ever increasing vastness
And dispersion of the very large—
It all going away, in a sense,
As well as
The ever decreasing compactness
Of the very small,
It, too, seeming to vanish;

Yet, we can neither see nor live
At either of these extremes
Because both of those paths lead
To very much the same state,
Which is well away from existence,
Going towards nonexistence,
Nothingness, even,
For there is only that
One alternative to existence—
The lack of anything.

So, it is, that,
Due to the one limiting case of the non,
The large is the same as the small—
The same vacant truth
On both ends of the size scale.

Illuminated Revelations

Illuminated Revelations

There is no ceiling to the universe,
And, just as importantly,
There is no floor.

The infinitely large and vast
Is too large to observe
And the infinitely small and compact
Is just as inaccessible.

Our finite existence lies suspended—
Well above the microscopic world
Of the infinitely small
And well below the immensity
Of the infinitely large.

What keeps us hanging there?

Why does our reality not shrink
Into a single point?

Why this location
On the cosmic size scale?

...

It is because
We are the singularity of existence,
Perched here between
The infinitely large and small,
The only place we can be,
Halfway between
Infinite largeness
And infinite smallness
Because they are the same thing.

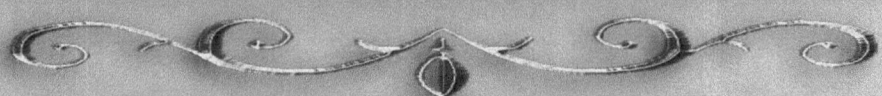

Illuminated Revelations

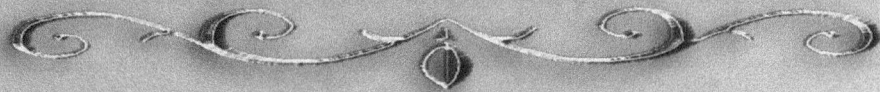

Illuminated Revelations

It's no wonder, then,
Why zero and infinity
Cause some of the same problems in math,
For they are two different viewpoints
Of the same thing…

And so, too, is this always reflected
In the problems that particle physicists
Have in trying to find a connection
Between the macro and micro universe.

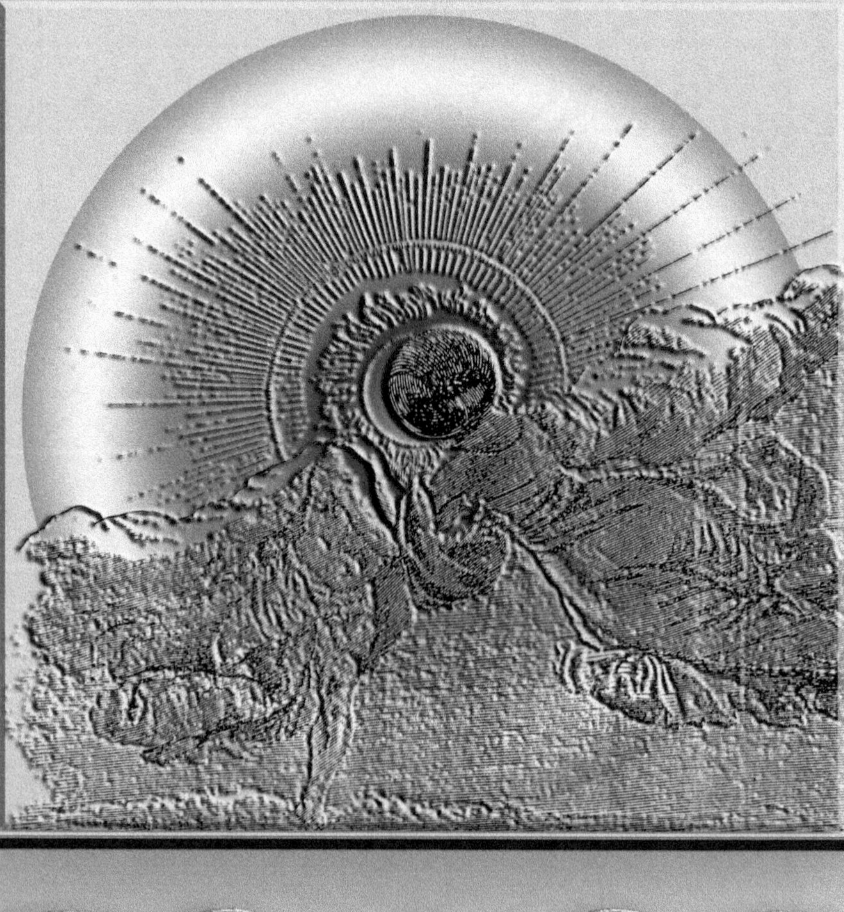

Illuminated Revelations

Well, we have already come far
In this essay,
Even to the end, in a way,
If you have absorbed these ideas
To see where we have gone,
But, the explanations will continue.

The Self-Referencing Reality
And its "Vanishing"

Our reality is self-referencing,
For we have lengths
Of varying finite sizes—
A clear order to the size of things,
But for the infinite extremes;
However, this is because finite size
Is the only size available
For us to measure,
Mired, as we are and must be,
At the midpoint
Of the existential scale.

With no ceiling and no floor
We are a mean between
Nothing and everything,
Each having
No real information content.

We are a Nothing
Compared to the Infinite,
Although an All
In comparison with the Nothing;
Yet, All and Nothing
Are synonymous perspectives
Of the same thing;
For nothingness is the only form
That the completeness
Of symmetry can take...

...The only place to go from finiteness,
The only place to go from existence.

Illuminated Revelations

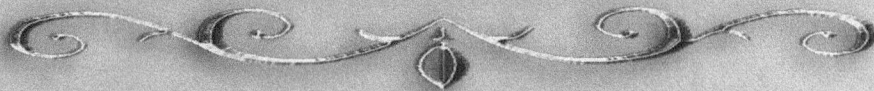

All that a point in space
Has going for it is
Its position relative
To other points,
For it has
No intrinsic
Properties.

From the "outside"
It all vanishes,
Like the grin
On the
Cat.

(Cheshire Cat)

A superposition of any number of points
At the same position
Is indistinguishable from a single point;
Thus, the multitude is the same as the one.

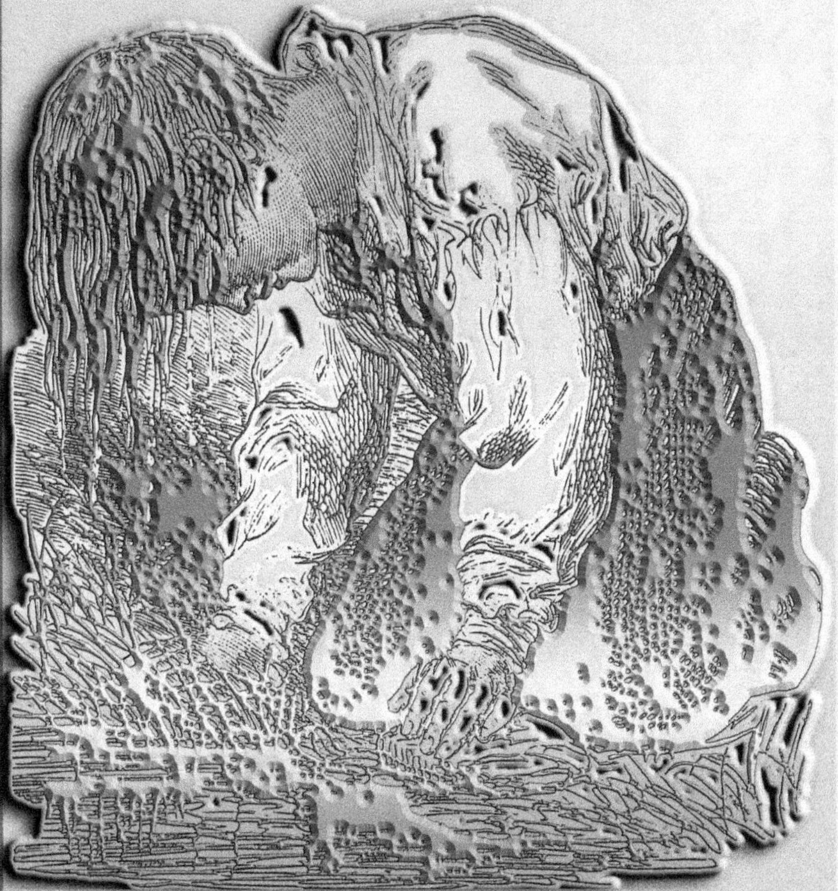

There can be no begininngs or ends
To the universe
In the eternally perfect symmetry.

Illuminated Revelations

We exist in the everlasting universe
In the finite middle of nowhere
Between the two infinities,
The reality upon the symmetry of forever.

"Live, for the rose withers all too Soon!"

Illuminated Revelations

The Moving Finger writes; and, having writ,
Moves on: nor all thy Piety nor Wit
Shall lure it back to cancel half a Line,
Nor all thy Tears wash out a Word of it.

The Perfect Symmetry

Particles and antiparticles
Are always created in pairs.

50% of the universe is antimatter,
Which is the ultimate supersymmetry.

A particle and an antiparticle may collide;
Photons are then emitted in opposite directions
To preserve and conserve momentum.

These photons may eventually
Play a role in the construction
Of another particle and an antiparticle.

Matter is necessary to create light
And light is necessary to create matter.

Photons have no charge
And so they are neutral,
But it's really that they
Carry a positive
And a negative charge
That sums to neutrality.

Particles and photons,
Whether existing or created,
Cannot do other than they do,
For they are bound to symmetry;
Determinism is the result.

So, there is no mystical statistical
Quality of quantum randomness;

Illuminated Revelations

Illuminated Revelations

It's all just the universe-wide,
Perfect equation remaining in balance,
All fields stretching forever
And ever affecting one another.

A universe is not just big contraption
That has to follow a few laws,
But is an utterly flawless system
That is fully defined and determinant.

There can be no deviation.

Each electron and photon
Have the exact momentum
They are supposed to have,
To infinitely fine resolution,
An infinite history of energy.

There can be no modification.

A particle is not a point
But has an infinite spread,
Although going to zero at infinity;
So, its creation "wave"
Is followed at some distance
By its annihilation "wave".

Particle creation/annihilation, though,
Is not the main recycling event,
But the ongoing galactic fusion is—
The reshuffling and recycling
Of the atomic elements.

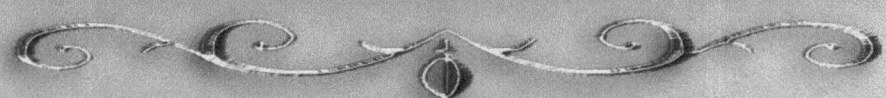

Illuminated Revelations

Illuminated Revelations

The overall universe never changes,
For the amount of energy stays the same,
Nor does the amount of charge change,
For there are infinite number of
Positive and negative electric fields.

There is no problem in having
An infinite amount of positive
And negative energies and charges,
Since, due to the perfect symmetrical equation,
They, like anything else, cancel to zero,
For existence itself must sum to nonexistence.

So, space is everywhere filled
With fields of energy;
Space and its distortions of matter
Are all that there is.

Reality pulsates,
In its real and structured sequence,
A field that's present throughout space immense,
Out of which all particles can condense—
Occurring where the field's extremely intense.

Particles are those bundles of inertia,
The knots in the field and fabric of space;
Yet, matter defines the structure of space…
So the Yin is in the Yang, and vice-versa!

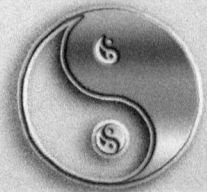

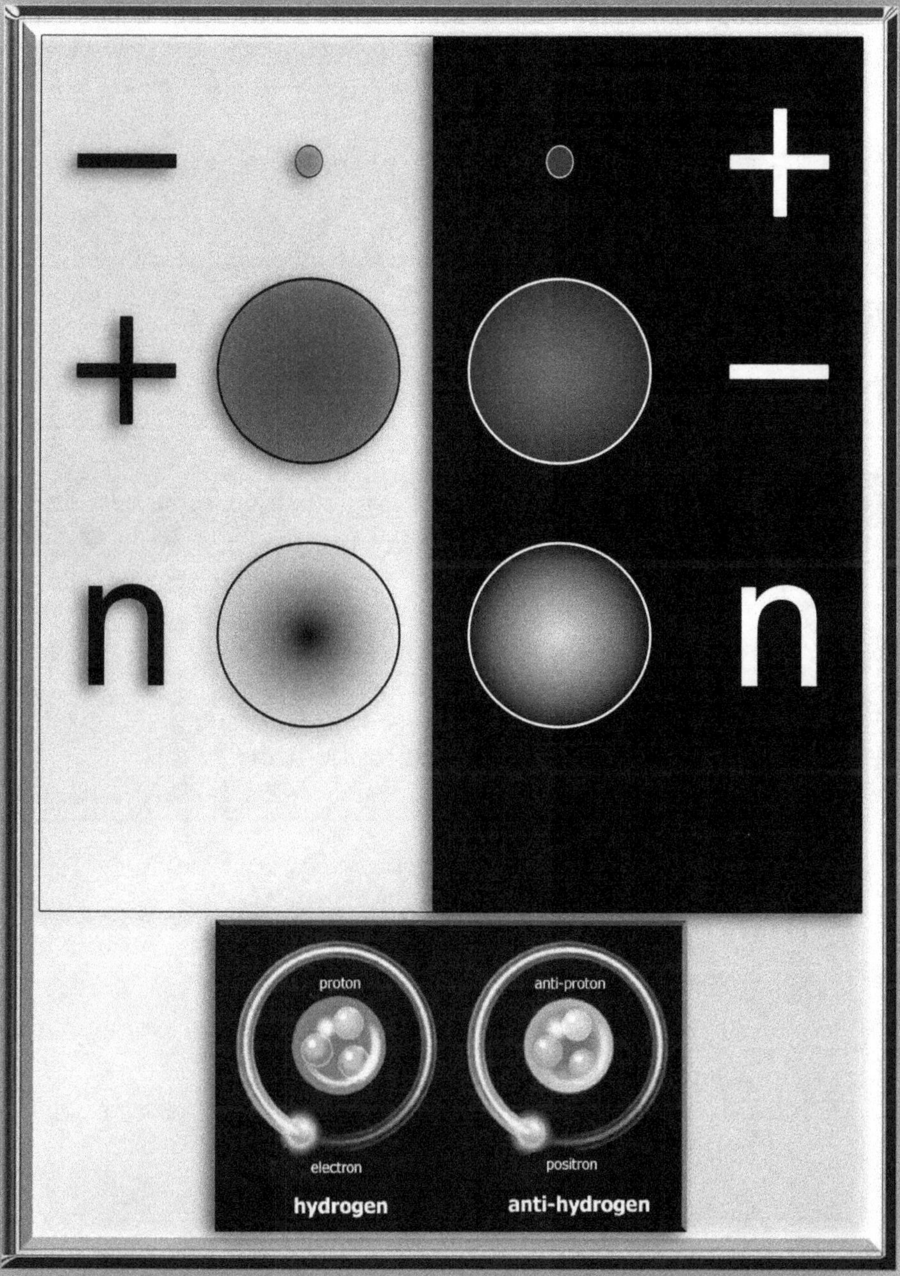

Illuminated Revelations

−   ·

+

n

+

−

·

n

proton   anti-proton

electron   positron

hydrogen   **anti-hydrogen**

Surface Boundaries

A 2D circle
Has a 1D surface boundary—
A line of infinite thinness
That is unbounded since
It goes 'round and 'round forever.

A surface boundary is always
One dimension less than
What it bounds.

A 3D sphere is bounded by
A boundless, infinitely thin 2D area
Which can be endlessly traversed.

It would seem to follow, then,
That a 4D hypercube
Would be bounded by
Our unbounded, infinitely thin 3D space,
Making this 4D hypercube finite,
Although our 3D surface space is endless.

So, this makes space a bounding surface,
Not a bounded interior region,
But still the boundary surface
Of its own totality.

The finite hypercube makes for
And defines an absolute length scale,
Which is why atoms behave differently from galaxies
And is how the quantization of energy is governed.

We can derive its 4D distance**4 quality.

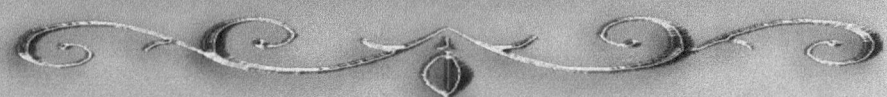

Illuminated Revelations

The Hypercube

The finite hypercube makes for
And defines an absolute length scale,
Which is why atoms behave differently from galaxies
And is how the quantization of energy is governed,
For the hypercube is the boundary condition—
And the only one—for energy quantization.

Energy is both contained and packaged
By the limitations of space.

Unit hypervolume is why it is possible
To have a universal constant
For microscopic processes
Like the creation of particles
And emission of photons.

The 4D hypervolume
Boundary condition's 4D constant
Is the physical manifestation
Of unit hypervolume,
And is proportional to the value

Hypervolume = C (hc/P) = C (E,w)/P

Which is the product of Planck's constant
And the speed of light
Divided by average universal energy density,

Where 'C' is the 4D constant,

'h' is Planck's constant
(In units of energy-time),

Illuminated Revelations

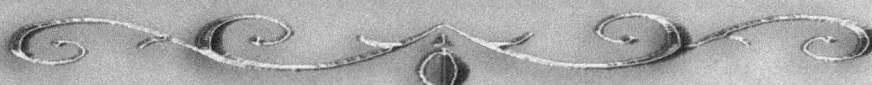

Illuminated Revelations

'c' is the speed of light
(In units of distance/time),

'E' is energy
(In units of time-distance**2)

P is universal energy density,
(In units of energy/distance**3
Or time/distance
Since Energy/Volume = tdd/ddd = t/d)

And 'w' is wavelength
(In units of distance).

Planck's constant, h,
Defines the size of quanta,
And is 4D because energy is 3D,
It being spread throughout space
In the form of an endless
And variable density distribution.

So, h = E/f

Where 'E' is the wave energy
And 'f' is the wave frequency.

Also, E = hf
(In units of time-distance**2)

Wavelength (w) and frequency (f)
Are related to the speed of light (c) as

c = wf
(Being in units of distance/time)

Illuminated Revelations

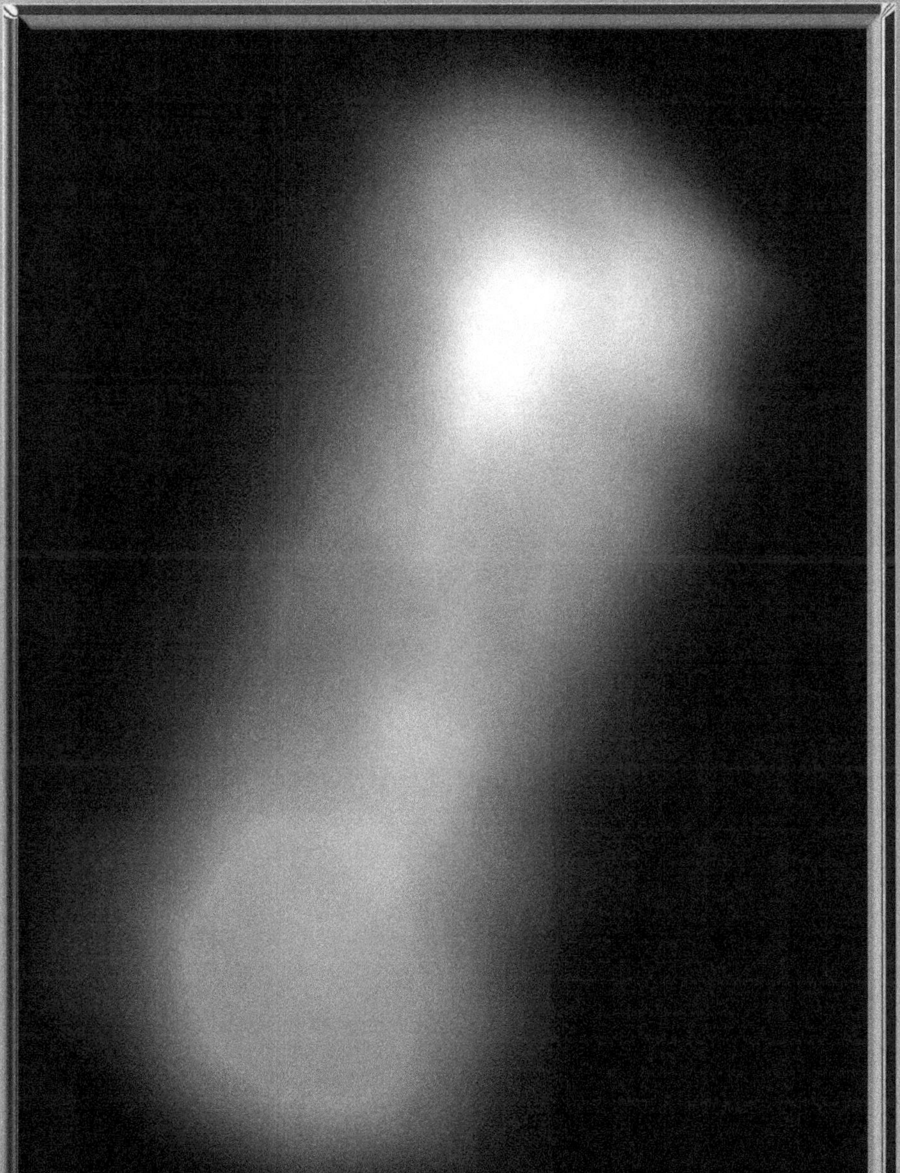

Illuminated Revelations

Thus,

$$h = E/f$$

And

$$c = wf$$

combine into

$$hc = Ewf/f \quad \text{or} \quad hc = Ew$$
(In units of time-distance**3).

We can derive the 4D constant,
'C', of proportionality,
That is a constant defined
By the way the energy
Of individual photons
Is geometrically distributed into space,
A manifestation of unit hypercube

$$\text{Hypercube volume} = C\,(hc/P) = C\,(Ew/P)$$

which, is, in units, for (hc)/P is

$$(\text{time-distance**3}) \ / \ (\text{time/distance})$$

$$= \text{distance**4}$$

So, the 4th dimension is one of closure.

Eternity is long enough for an infinite amount
Of matter/energy to express
Every possible spatial distribution.

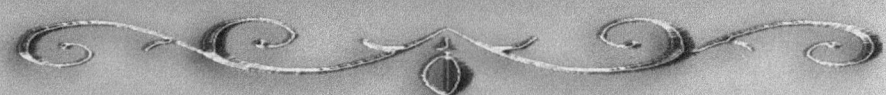

Illuminated Revelations

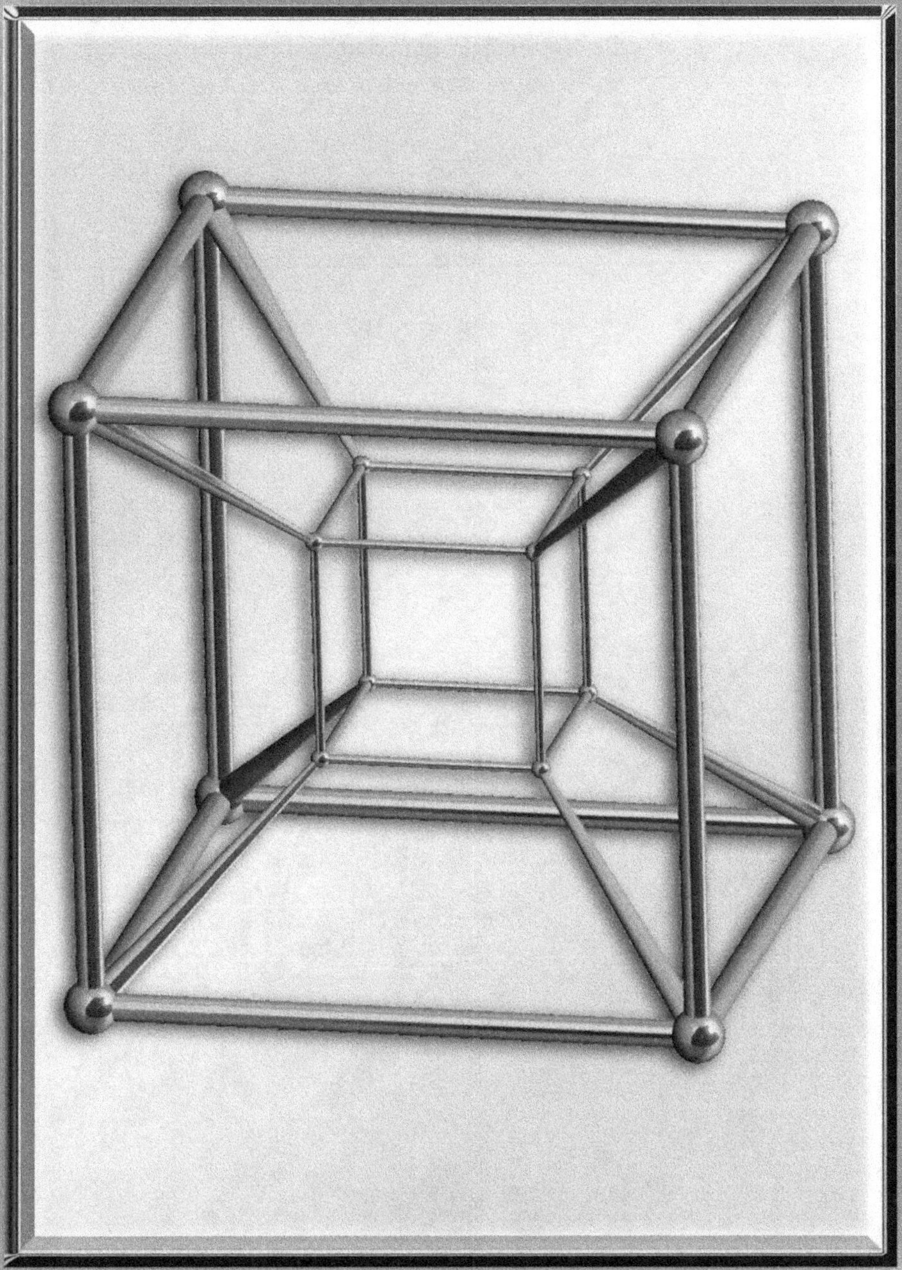

Illuminated Revelations

The two 3-D quantities
Of 4D hypervolume are

distance**3
(Space)

And

time-distance**2
(Energy)

Energy moves through space.

Illuminated Revelations

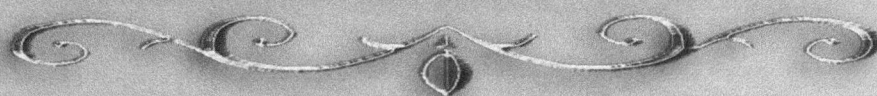

Illuminated Revelations

The space of our universe
Is three-dimensional
Because this is
The only dimension
Whose volume is
Compositionally consistent
Through all levels of infinite size
While forming the surface
Of its own hypervolume.

Time is the dimension that bounds,
Not extends,
Three-dimensional space.

Just as time is the difference of space,
So, too, is space the difference of time,

Unit hypervolume is the internal product
Of time and space,
But is also the product
Of energy and distance.

The speed of light is the dimensional equivalent
Between space and time.

Energy density is the 4th-dimensional slope of space.

Just as Planck's constant is
The four-dimensional quantization of photons,
Elementary charge is the four-dimensional
Quantization of particles

Photons are the encapsulation of time by space;
Particle fields are the encapsulation of space by time.

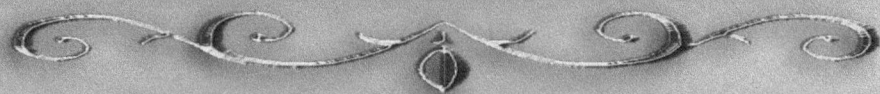

Illuminated Revelations

Time

Time is the difference of space
And space is the difference of time;
So, time is a difference dimension,
Not a compositional dimension.

Its two forms are the displacement
Caused by motion
And the polarity of electric fields.

The universe's totality
Is neutral and symmetric,
Whereas its internal composition
Is polar and asymmetric.

zero + zero = zero
(For spatial continuity,
The sum of nothing being nothing))

and

zero - zero = zero
(For polarity,
The difference of nothing being nothing)

Hypercube:

```
            C
            h
     S p a  c  e
            r
            g
            e
```

Illuminated Revelations

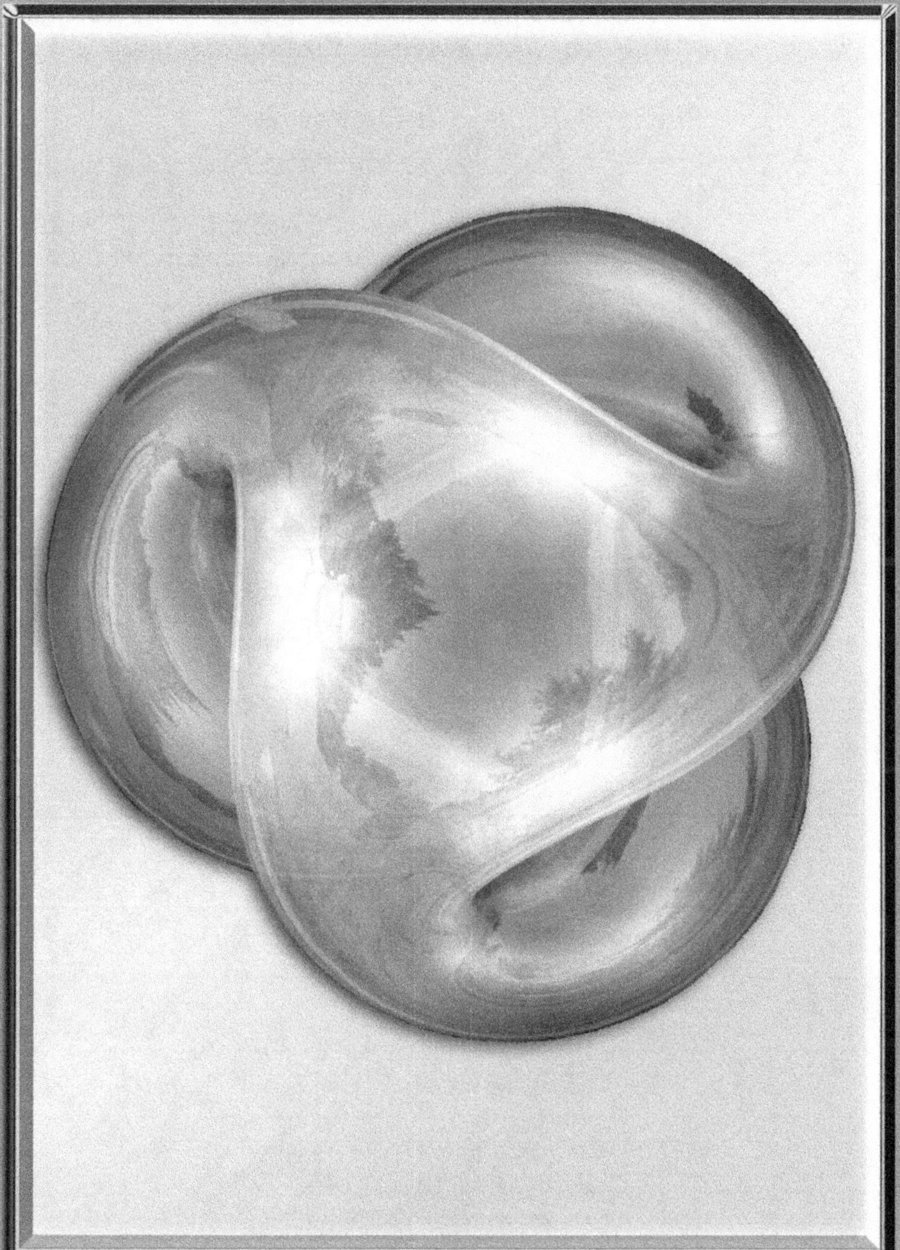

Polarity and Electric Fields

The universe is four-dimensional,
Its three infinite spatial extents
Defined by summation,
Leaving only one degree of freedom
For the difference operator
To perform the nullification of space:
The fourth dimension of polarity (time)
That has a positive and negative axis.

(+)

E
n
S p a c e (neutral)
r
g
y

(-)

The fourth dimension is not composed of points
Because it represents their 4-dimensional deflection—
A difference of position.

Distance***4 = c(time-distance**3)

'c', the speed of light,
Underlies the dimensional relationship
Between time and distance.

There is unbounded duration
Over infinite distance—
An eternity of infinity!

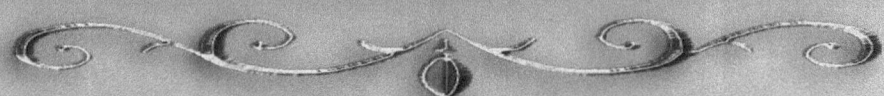

Illuminated Revelations

Recycling the Universe

Red-shift is due to the expansion
Of photons due to their decay,
Yet the CMB does not decay
Due to its deactivation
By its thermalization
With intergalactic material.

Optical photons reconcile
The energy they have lost
From intergalactic red-shift
By emitting microwaves directly
Into the CMB band.

A galaxy's halo pulls power Out of the CMB
With huge, deep space electrical currents
That leave slight temperature ripples in their wake.

The currents pass through the galaxy's disc
Toward its core, providing power
For the disassociation of
The compound nuclei created by fusion,
Liberating hydrogen,
Which flows up the galaxy's arms
And is ejected from its core
Directly into space.

Galaxies are vortices,
And so they carry the material
Of their disc region slowly into their core
Where it is absorbed into
A gravationally-expanded neutron superfluid
That exudes hydrogen…

Illuminated Revelations

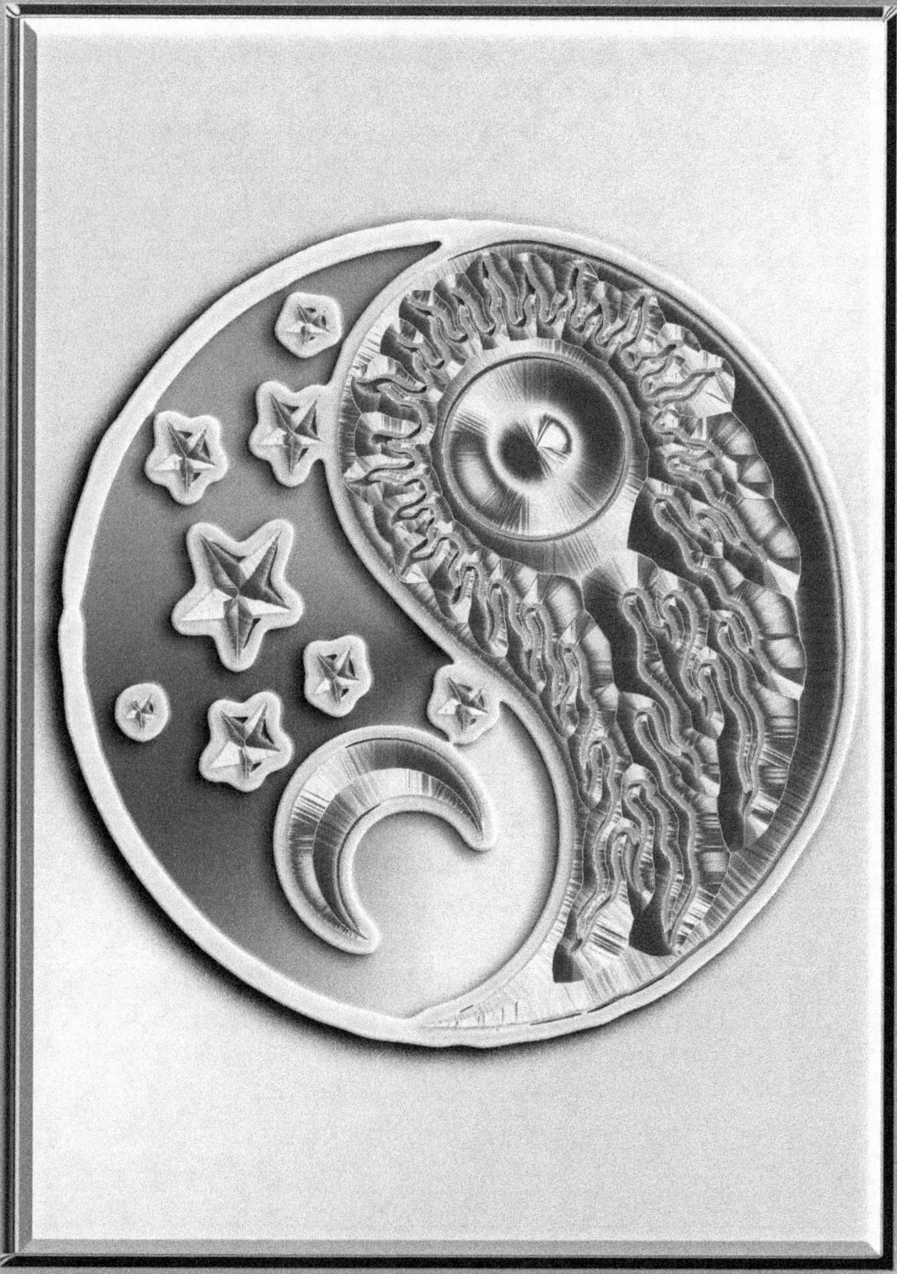

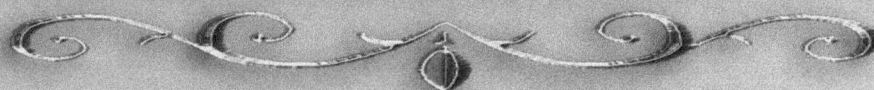

Illuminated Revelations

...Through its degenerate surface,
Thus providing a renewable source
Of fuel for the perpetual cosmic engine.

All true galaxies have massive,
Electrically charged black holes
In the innermost depths
Of their central region,
A necessary and integral component
Of galactic function because they
Are the only objects in the universe
With a gravitational potential large enough
To provide an environment capable
Of low-temperature nuclear disassociation
With virtually no radioactive energy loss.
(Without them the cosmic fusion cycle
Would not be possible.)

The reason our universe
Appears To be 12-18 billion years old
Is because this is the average time
Required for material to cycle
Through its galactic systems.

The universe is infinitely older than this,
But the compound nuclei of which
Its luminous material is composed
Are continuously renewed
Every 12-18 billions years.

This is cosmic equilibrium.

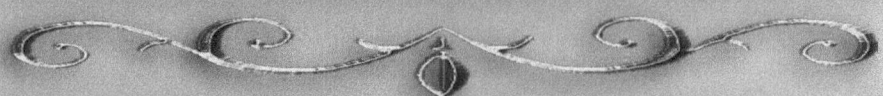

Illuminated Revelations

Illuminated Revelations

The Riddle of Existence

symmetry
asymmetry

Illuminated Revelations

Symmetry, ever symmetry
And forever symmetry forever.

Matter and light each make the other,
Both having always been—
The egg and the chicken
Making/unmaking at the same time.

The atomic elements are bundled in the stars,
Then dismantled in the galactic cores.

There are only two stable particles,
The electron and the proton,
Oppositely charged,
(Along with their mirror twins)
Because pair production
Only has two states
Able to generate separate particles.
(All other states collapse.)

The All and the Null
Are each other's echo.

More is less and less is more—
The largest and the smallest being the same.

The Great Question:

Of what else could existence be made
When there is nothing prior?

Illuminated Revelations

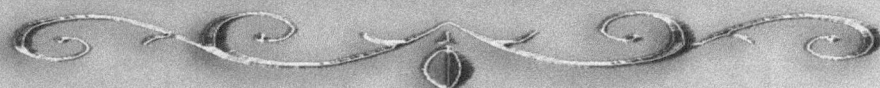

The answer is right there in the question:
Nothing, for there can be no alternative.

It is not that existence
Came from
Nonexistence or Nothing,
But that it _is_ Nothing—
And so existence ever was and is
Because Nothing ever is and was.

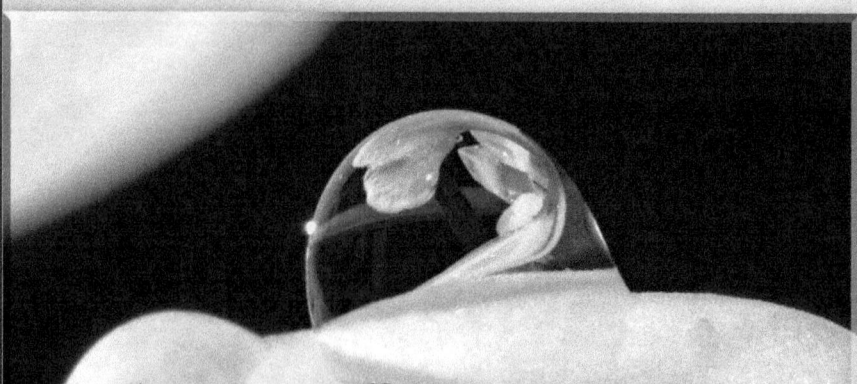

To see a world in a grain of sand,
And a heaven in a wild flower,
Hold infinity in the palm of your hand,
And eternity in an hour.

Regret <—— (ambigram) ——> Nothing

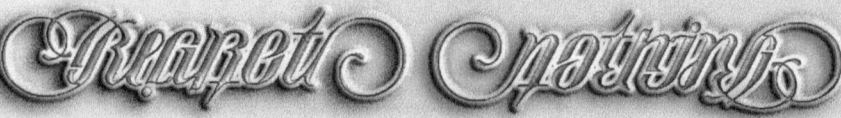

Illuminated Revelations

Nada   Nothing

Nought   Zero

$$1 + (-1) = 0$$

Illuminated Revelations

We are the lustrous and glowing arc
Of reality's scintillating rainbow
(And it's gleaming reflection, too)
That spans the symmetric infinity,
Glistening here, on forever's edge of fixity.

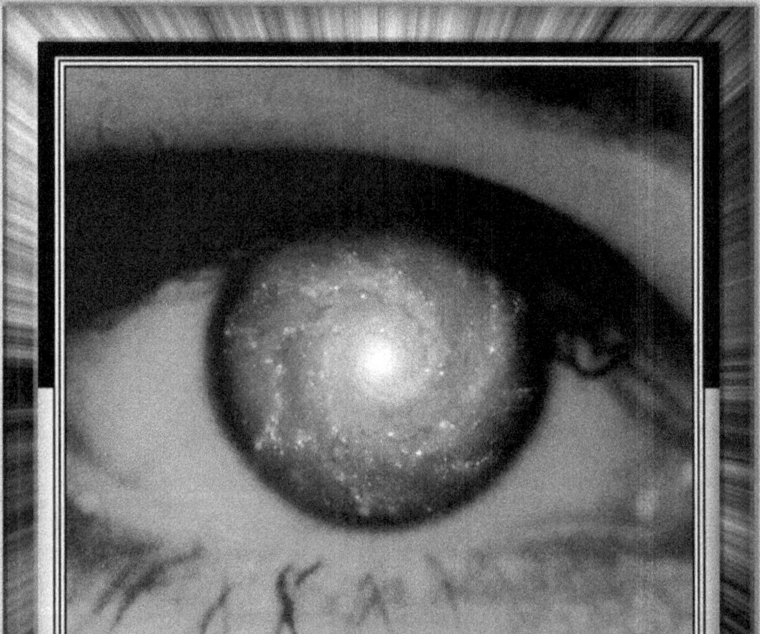

Illuminated Revelations

Q. E. D

(Nobody Nowhere)

Being
Nothingness

Our parentheses in eternity
Flashes as a twinkling, but's extended
By time into a phantasmic life dream
That's existent the same as if it were.

A life dream's like a rainbow, not really there,
A false phenomenon become tangible
Through its being, the true true of the faux true,
Molding a genuine significance.

Life's indeterminate or not, the same
Being brought by the virtual as the true,
The mechanics being as incidental
As why "primary color" chose its waves.

Life's here, like a virtual particle
Born this side of an event horizon
Of a Black Hole, realized by its presence
In the realm of what's been radiated.

There is no difference of what makes none;
Realism is now playing, the living film:
A reality show in the theater
Of the mind's eye, with the 'I' observing.

www.ingramcontent.com/pod-product-compliance
Lightning Source LLC
Chambersburg PA
CBHW071410170526
45165CB00001B/228